SIMPLE ACTIONS for Jews to Help GREEN the Planet
JEWS, JUDAISM AND THE ENVIRONMENT

Rabbi Dov Peretz Elkins
Co-author, CHICKEN SOUP FOR THE JEWISH SOUL
Winner of National Jewish Book Award
FORWARD by Rabbi David Saperstein

Copyright © 2011 Dov Peretz Elkins
All rights reserved.

ISBN: 1463777655
ISBN-13: 9781463777654
Library of Congress Control Number: 2011913720

CreateSpace, Charleston, South Carolina

Praise for *Simple Actions for Jews to Help Green the Planet*

In this important book, Rabbi Dov Peretz Elkins reminds us all of the perils that our environment faces today, and he specifically challenges us to look to our heritage as a guide to becoming better stewards of our earth. As God commands Adam in the Book of Genesis to protect the garden, Rabbi Elkins too challenges us to be protectors of God's good earth and everything that exists on it.

Senator Joseph Lieberman

This is an extraordinarily practical guide to what we all could and should be doing to reduce our own footprint in the world and to help live a commitment to the survival of our planet, and it is beautifully rooted in Jewish text and teaching. Read and ACT; lots of small steps will help make big change.

Ruth W. Messinger, President, American Jewish World Service

"We shall do – and We shall listen" meets "Think Globally Act Locally": Rabbi Elkins' thoughtful and useful volume -- with its rare combination of traditional wisdom and very practical, green, lifestyle recommendations -- is required reading for anyone who cares about Judaism and environmentalism.

Professor Alon Tal, Ben Gurion University -- Founder of the Israel Union for Environmental Defense and the Arava Institute for Environmental Studies

---

This thoughtful guide to 21st century Jewish environmental living provides the context for contemporary Jewish eco-consciousness with a useful focus on how to actually walk the talk!

Sybil Sanchez Director - COEJL - Coalition on the Environment and Jewish Life Jewish Council for Public Affairs

---

With humanity now a significant factor in the future of life on Earth, we must mobilize to live wisely and to steward this green/blue jewel of a home. Dov Peretz Elkins is our sage guide - showing us a path to sanity, serenity, and beauty.

Rabbi Bradley Shavit Artson, Abner and Roslyn Goldstine Dean's Chair, Vice President, American Jewish University, Ziegler School of Rabbinic Studies

Dov Peretz Elkins has created a wonderful Jewish environmental action guide. When I speak about the environment, I am often asked "What can I do?" Here are all the answers and every synagogue Green Team and Jewish home can find many different ways in this book to help to bring about *Tikkun Olam*.

Rabbi Lawrence Troster, Rabbinic Scholar-in-Residence, GreenFaith, and environmental theologian and activist.

Jews understand from the story of Creation that people are the stewards of our planet. At a time of global environmental crisis, this book is laden with simple, straightforward ways that each of us can participate in keeping our world green and habitable. It guides us gently into actions that we can take in our daily lives to make a difference for our planet.

Rabbi David A. Teutsch, PhD. Wiener Professor of Contemporary Jewish Civilization and former president, Reconstructionist Rabbinical College

---

Finally! A big, green Jewish guide to how to live sustainably on this increasingly fragile planet of ours. There is much wisdom in the Jewish tradition for how to do this, and Rabbi Elkins is a knowledgeable and experienced teacher. He makes the much-needed connections between learning and life, in a way that is accessible, relevant and compelling. This book is a real contribution to an important and growing field, and should find its place in every Jewish school and home.

Dr. Jeremy Benstein, deputy director of the Heschel Center for Environmental Learning and Leadership, Israel, and author of The Way Into Judaism and the Environment *(Jewish Lights, 2006).*

Yishar Kohakha to Rabbi Elkins for this important and timely collection of many important insights about the Judaism and the environment from our ancient past to our current challenges. Every chapter has recommended activities to lead a Jew from theoretical interest in the environment to sustainable behaviors that will make the world a better place for all humanity. Read this book, let it guide the changes you need to make in your life, and share it with others.

Rabbi Eric M. Lankin, DMin Chief, Institutional Advancement and Education, Jewish National Fund

Rabbi Elkins provides practical actions embedded in a wealth of thought- from the mystics of the ancients through to the scientific of the modern, to encourage our personal paths of harmonizing with the Earth's needs and receiving the blessing of Life.

Rav Shaul David Judelman, a Seattle native, received his rabbinic ordination from the Bat Ayin Yeshiva and currently leads tours and programs in Israel combining learning and action around the theme of Jewish Ecology. He can be reached at organicjew@yahoo.com and is full of ideas as to how to come to Israel in a Green way.

Dov Peretz Elkins' new book is comprehensive and insightful. It is a welcome update on the existing body of work on the Jewish environmental movement. I recommend that every Jewish family, educator, and community add this to their reading library.

Nili Simhai, Co-Director, Teva Learning Alliance

This book combines usable and relevant information with a strong Jewish frame. It's well-written, it's accessible and it provides a slew of practical ways that you can live more sustainably. You should buy a copy - and you should use it!

Nigel Savage, Founder, Hazon: Jewish Inspiration, Sustainable Communities

---

This is the most complete and comprehensive treatment of Judaism and the environment, with special emphasis on actions. A user-friendly how-to book, and a quick-read.

Rabbi Charles Simon, Executive Director, Federation of Jewish Men's Clubs, and author, Building a Successful Volunteer Culture *(Jewish Lights Publishing)*

Simple Actions is far more than just another list of ways to "green" the planet. Rabbi Elkins has provided us with a resource long overdue – a single book that includes ancient Jewish environmental teachings; contemporary voices of the Jewish environmental movement; and practical and MEANINGFUL ways for Jews of all ages and from all walks of life to live more environmentally conscious lives. Thank you Rabbi Elkins!

Dr. Gabe Goldman - Director of Jewish Experiential and Nature Education, American Jewish University

Do you think that Judaism has nothing to do with environmentalism? Read this book and think again: Rabbi Dov Peretz Elkins presents a comprehensive how-to guide to discovering and practicing Judaism at its very best — environmentally sustainable, as God intended.

David Krantz, President and Chairperson of the Green Zionist Alliance:
The Grassroots Campaign for a Sustainable Israel

BOOKS BY DOV PERETZ ELKINS

See: http://www.jewishgrowth.org/cgi-bin/books.cgi?action=catalog

Simple Actions for Jewish to Help Green the Planet
Jewish Stories From Heaven and Earth
Chicken Soup for the Jewish Soul
The Wisdom of Judaism: An Introduction to the Values of the Talmud
Rosh Hashanah Readings
Yom Kippur Readings
The Bible's Top 50 Ideas (with Abigail Treu)
A Shabbat Reader
Moments of Transcendence
Jewish Guided Imagery
Forty Days of Transformation
Meditations for the Days of Awe
Shepherd of Jerusalem
Moments of the Spirit: Quotations to Inspire, Inform & Involve
Enveloped in Light: A Tallit Sourcebook
A Treasury of Israel and Zionism
The Eulogy Book
Prescription for a Long and Happy Life
A Tradition Reborn: Sermons and Essays
My Seventy-Two Friends: Encounters with Refuseniks in the USSR
God's Warriors: Dramatic Adventures of Rabbis in Uniform
Clarifying Jewish Values
Jewish Consciousness Raising
Experiential Programs for Jewish Groups
Teaching People to Love Themselves
Glad To Be Me: Building Self-Esteem in Yourself and Others
Twelve Pathways to Feeling Better About Yourself

TRANSLATIONS
(original Hebrew books by Simcha Raz)
Hasidic Wisdom (with Jonathan Elkins)
The Torah's Seventy Faces
Tales of the Righteous

BOOKS FOR TEENS
(both books with Azriel Eisenberg)
Worlds Lost and Found: Discoveries in Biblical Archeology
Treasures From the Dust: Biblical Archeology

CHILDREN'S BOOK
Seven Delightful Stories for Every Day (ages 3 to 6)

...God said, "Look at My works. See how beautiful they are, how excellent! For your sake I created them all. See to it that you do not spoil or destroy My world - for if you do, there will be no one to repair it after you."

Midrash Kohelet Rabbah 7.13

Even those Creatures that you deem superfluous in this world, such as flies, fleas and gnats, nevertheless have their allotted task in the scheme of Creation.

Midrash Shemot Rabbah 10:1

Any intelligent fool can make things bigger, more complex, and more violent. It takes a touch of genius – and a lot of courage – to move in the opposite direction.

Albert Einstein

If we could change ourselves, the tendencies in the world would also change. As a man changes his own nature, so does the attitude of the world change towards him. ... We need not wait to see what others do.

Mahatma Gandhi

Hope for the earth lies not with leaders but in your own heart and soul. If you decide to save the earth, it will be saved. Each person can be as powerful as the most powerful person who ever lived – and that is you, if you love this planet.

Helen Caldicott

To Yoni and Pesach Jeremy —
first my sons, then my students, now my teachers

Contents

Introduction

Figure #1: Statement on Climate Change

Section A: The Basics.....................1

Number One- Think Global: Measure and Reduce Your Ecological Footprint

"The Heavens declare the Glory of God; Of God's handiwork the sky boasts… Their shout rings throughout the earth; their words to the end of the world."   Psalms 19:2-5

Number Two- Act Local: Get to Know Your Local Environment

"The people of Israel is like a walnut pile. If one walnut is removed, each and every walnut in the pile will be shaken." *Midrash Song of Songs Rabbah* 6:26

Number Three- What's a Green Jew? Learn what the Torah says about Protecting the Environment

"The earth belongs to Ado-nai, and all that it holds; the world and its inhabitants." Psalms 24:1

"Rabbi Shimon bar Yohai said: 'The shade spread over us by these trees is so pleasant! We must crown this place with words of Torah!'" Zohar

"…destroying a lot of little species matters to the same extent that taking out a lot of little screws from an airplane matters." Pulitzer-prize winning scientist Jared Diamond

Figure #2: Ten Teachings on Judaism and Environmentalism

Figure #3: Jewish Texts as Resources on Global Climate Change
Figure #4: Pillars of Jewish Environmental Awareness

Section B: So Many Ways to Save Energy 23

Number Four- Home Energy Use

"Make the planks for the Tabernacle of acacia wood...." Exodus 26:15. Why of acacia wood? God set an example for all time, that when we are about to build a house from a food tree, we should be reminded: If when God commanded the Sanctuary to be erected, we were to use only barren trees. Even though all things belong to God – how much more should this be so with wood for our own house. *Midrash Exodus Rabbah* 35:2

Number Five- Get an Energy Audit

"If a person kills a tree before its time, it is like having murdered a soul." Rebbe Nahman of Breslov (d. 1810)

Number Six- Green Energy

"The sun and moon stand still on high; by the light of Your arrows do they go, by the illumination of Your shining spear." Habakkuk 3:11

And the sun says:

"Arise, shine, for your light has dawned! The Presence of Ado-nai will shine upon you!" Isaiah 60:1

Number Seven- Green Your Transportation

"I will sing to Ado-nai, for God has triumphed gloriously; horse and driver God has hurled into the sea." Exodus 15:1

Number Eight- Use Carbon Offsets

"God hoped for justice, but behold, injustice. For equity, but behold, iniquity! Woe to those who add house to house, and join field to field, till there is room for none but you to dwell in the land." Isaiah 5:7-8

Section C: Bal Tashkhit: Reduce Waste 43

Number Nine – Avoid Disposables: Cups and Plates and Much More

> "Not only one who cuts down food trees, but also one who smashes household goods, tears clothes, demolishes a building, stops up a spring, or destroys food on purpose violates the command: 'You must not destroy!' (Deuteronomy 20:19)." Maimonides

Number Ten – Reducing the Impact of Electronics Waste

> "The purposeless destruction of anything at all is forbidden. ... Only for wise use has God laid the world at our feet, when God said to humankind, 'Fill the earth and master it, and rule!'" (Rabbi Samson Raphael Hirsch – d. 1888)

Number Eleven – Drink Tap Water Instead of Bottled Water

> "Waters shall burst forth in the desert, streams in the wilderness.... Parched land shall become fountains of water." Isaiah 35: 6-7

Section D: Reduce, Reuse, Recycle 57

Number Twelve – Reduce

> "The more possessions, the more worries." Pirke Avot 2:8

Figure #5: Happiness: No Purchase Necessary

Figure #6: Simplicity as a Jewish Path

Number Thirteen – Reuse

> "All streams flow into the sea, yet the sea is never full. To the place from which the water flows, there it returns." Ecclesiastes 1:7

Number Fourteen – Recycle

> "Why did God choose a thorn bush from which to speak to Moses?... (Exodus 3:1-6). To teach you that no place is devoid of the divine Presence, not even a thorn bush." *Midrash Rabbah Exodus* 2:5.

Section E: Buy Sustainably. 77

Number Fifteen – Be a Conscious Consumer

"Not everyone engaged in commerce learns wisdom." Pirke Avot 2:6

Number Sixteen – Give Eco-Friendly Gifts and Greeting Cards

"Then those who fear God communicated with each other." Malachi 3:16

Number Seventeen – Eco-Friendly Tourism

"Then Ado-nai said to Moses, 'Why do you cry out to Me? Tell the Israelites to travel!'" Exodus 14:15

Section F: Connect with Creation. 87

Figure #7: Contemplating the Days of Creation

"The Heavens declare the glory of God; the sky proclaims God's handiwork." Psalms 19:2

Number Eighteen – Spend More Time Outdoors

"Shout, O heavens, for God has acted; shout aloud, O depths of the earth! Shout for joy, O mountains, O forests with all your trees!" Isaiah 44:23

Number Nineteen - Learn and Teach the Jewish Prayers of Thanksgiving for the World

"Rabbi Akiva taught: It is forbidden to enjoy anything in this world without first reciting a blessing…." Talmud, B'rakhot 35a

Number Twenty – Observe Shabbat

"More than the Jewish People has kept the Sabbath, the Sabbath has kept them." Ahad HaAm (d. 1927)

Section G: Greening Your Home 113

Figure #8: Protect the Environment: At Home and in the Garden

Number Twenty-one - Clean the Green Way

> "Who may ascend the mountain of Ado-nai? One who has clean hands and a pure heart." Psalms 24:3-4

Figure #9: Top 10 Eco-Friendly Ways to Clean the House

> "Rabbi Pinhas ben Yair taught: … Purity leads to holiness…." Talmud, Avodah Zarah 20b

Number Twenty-two - Use Your Bathroom Wisely

> "Abayei said, when one comes out of a privy he should say: Blessed is God Who has formed us in wisdom and created in us many orifices and many cavities. It is obvious and known before Your throne of glory that if one of them were to be ruptured or one of them blocked, it would be impossible for a human to survive and stand before You. Blessed are You that heals all flesh and does wonders." Blessing after exiting the bathroom, Talmud, B'rakhot 60b

Number Twenty-three – Compost

> "Blessed are You, Ado-nai our God, Who brings forth bread from the earth." Traditional Jewish Blessing before eating bread

Section H: Sustainable and Jewish Eating . .133

Number Twenty-four – Eat Organic

> "The Lord God took the human and placed him in the Garden of Eden, to till it and tend it." Genesis, 2:15

Number Twenty-five – Buy Local

> It is forbidden to live in a town that does not have a green garden. Jerusalem Talmud, Kiddushin 4:12

Number Twenty-six - Eat Less (or No) Meat, Consider Becoming a Vegetarian

> Rabbi Isaac ha-Levi Herzog [first Chief Rabbi of Israel, d. 1959] said, "Jews will move increasingly to vegetarianism out of their own deepening knowledge of what their tradition commands.... A whole galaxy of central rabbinic and spiritual leaders...has been affirming vegetarianism as the ultimate meaning of Jewish moral teaching."

Figure #10: Why I became a Vegetarian

Number Twenty-seven – Sustainable Meat Option

Be it further resolved that our organization will evaluate Kosher food manufacturers in the areas of
- Employee welfare, including fair wages, benefits, health safety;
- Employee training;
- Quality control and animal welfare;
- Corporate accountability and integrity; and
- Environmental impact.

[Model resolution for Jewish organizations]

Number Twenty-eight – When You Eat Fish

> And God said to them, "Be fruitful and multiply and fill the earth and subdue it and have dominion over the fish of the sea and over the birds of the heavens and over every living thing that moves on the earth." Genesis 1:28

Section I: Be a Green Jew 155

Number Twenty-nine – Make Your Jewish Life Cycle Celebrations Green

> "If our mouths were filled with song like the sea, and our tongues with joyous song like the roar of the waves, our lips with praise like the expanse of heaven, and our eyes were to shine like the sun and the moon. If our hands were spread out like the wings of the eagle, and our feet were as swift as the hinds, it would not be enough to thank You, Ado-nai our God and God of our ancestors, and to bless Your Name for even one of the thousands and myriads of favors, miracles and wonders that You have done for us and our ancestors . . . ." Traditional Jewish Liturgy – Shaharit (Morning) Service

Figure #11: 7 Ways to Green a Bar/Bat Mitzvah

Figure #12: Rules for Greening a Wedding

Number Thirty - Green Your Jewish Holidays

"The holidays are the jewels on the crown of Judaism." (Rabbi Abraham J. Karp, *The Jewish Way of Life*).

Number Thirty-one – Celebrate Tu b'Shevat

Tu B'Shvat, the holiday of fruit trees, recalls the Garden of Eden, and the human quest for spiritual refinement. Rebbetzin Tzipporah Heller

Number Thirty-two – Plant Trees

"Just as my ancestors planted for me, so I plant for my descendants." Talmud, Ta'anit 23a

Section J: Build a Green Jewish Community.187

Number Thirty-three – Get to Know the Jewish Environmental Movement

"Separate not yourself from your community" Pirke Avot 2:5

Number Thirty-four - Help Your Synagogue and Other Jewish Institutions Go Green

"Rabbi Shimon Bar Yohai said,' three things are of equal importance, earth, humans, and rain.' Rabbi Levi ben Hiyyata said:...'to teach that without earth, there is no rain, and without rain, the earth cannot endure, and without either, humans cannot exist.'" *Midrash Genesis Rabbah* 13:3

Number Thirty-five - Serving Healthy Food in Jewish Institutions

"And you shall eat and you shall be satisfied. And you shall bless Ado-nai, your God, for the good land God has given you."(Deuteronomy 8:10) Birkat Hamazon - Grace after Meals

Number Thirty-six – Learn about and Protect Israel's Environment

> "For Ado-nai your God brings you into a good land, a land of brooks of water, of fountains and depths, springing forth in valleys and hills; a land of wheat and barley, and vines and fig-trees and pomegranates; a land of olive-trees and honey;" Deuteronomy 8:7-8

Figure #13: Articles about Israel's Environmental Innovations

Conclusion – Become a Jewish Environmental Activist

> "In a place where there are no worthy people, you be a worthy person." Pirke Avot 2:6

Acknowledgments

Credits

Appendix: National Jewish Environmental Organizations

Resources: Judaism and the Environment; Books and Websites

Notes

Note: The paper used in this book is from partially recycled post-consumer waste material. The printer is in the process of increasing the percentage of recycled paper in future books, including forthcoming editions of this book.

A portion of the profits from the sale of this book will be contributed to the Coalition on the Environment and Jewish Life (COEJL), and to the Teva Learning Alliance.

Foreword
Rabbi David Saperstein

If our glorious planet is not one of God's greatest gifts to humankind, then what is? The obligation to protect God's creation makes our environmental crisis one of the most religious issues of our time.

We are called to protect something real, something substantial entrusted to our care. "In the beginning, God created the heavens and earth.... God called the dry land Earth; and the gathering together of the waters called the Seas: and God saw that it was good."

Our unprecedented ability to travel, taking us across the globe – indeed the view of Earth from outer space – allows us to see with wonder and awe the magnificence of God's creation—our blue seas and skies, our green fields and white capped mountains. We, too, can see it all and say:

"It is good."

Yet we see vividly as well the peril and the damage being done by our greed, our ignorance, our indifference: global climate change, the destruction of our ozone layer, the pollution of our air and water, deforestation of much of the globe, the skyrocketing rate of loss of bio-diversity; the growing scarcity of water that strains stability in so many environmentally fragile areas of the globe, including the Middle East; and the growing impact of alarming rates of population growth threatening to overwhelm our limited resources.

So God's creation -- the earth we inhabit -- is in danger. The seas, the forests and the rivers, the soil and the air, the environmental integrity of cities and towns – all are in peril. Our sense of urgency grows and is mobilizing the Jewish community.

Why? Perhaps this sense of urgency resonates so deeply within our Jewish sensibility: the lesson of Noah, that indifference to wickedness can lead to destruction of the world; and the prescient warning of the famous midrash "All that I have created, have I created for you. If you destroy it, there will be none after you to make it right again."

These lessons are imprinted on us as we have learned through the bitterest of tragedies that the unthinkable can happen when good people sit idly by.

Across the globe, we see growing engagement of religious communities, including a plethora of effective groups in the Jewish communities of North America and Israel. Religious institutions are seeking to become more sustainable in their practices; religious voices are calling for sustainable economic, business and political policies.

The religious community has, above all, thrust before the conscience of the world the recognition that people are already suffering because of the environmental degradation our species has caused. For some, it has meant death from disease or poisoning; for others, chronic health problems, loss of the quality of life, or a real loss of economic opportunities.

At such a moment of urgency and peril, hope and opportunity, at a time when across the spectrum of the Jewish community synagogues and Jewish communal institutions seek guidance on what to do, along comes *SIMPLE ACTIONS FOR JEWS TO HELP GREEN THE PLANET*.

Rabbi Dov Peretz Elkins, one of American Jewry's most talented writers and teachers, once again captures the pulse of the moment -- this time our environmental challenges -- in ways that lift the spiritual and religious sensibilities of this issue addressed so deeply in the sources of our tradition.

SIMPLE ACTIONS provides a vivid and most useful guide for individuals and synagogues as they try to grab hold of this crisis in ways that allow individual Jews and Jewish institutions to make a real difference.

That is the power of this book – to root this crisis in the compelling moral and religious texts and traditions of our people; to provide a blueprint for how individual Jews and Jewish institutions can respond effectively to this crisis; to offer a powerful call, mobilizing our community to lead in the urgent steps humanity must take — and in doing so, to ensure that future generations, looking back in a world in which God's creation once again flourishes, will look at our generation and say: this our mothers and fathers did for us.

Rabbi David Saperstein is director of the Religious Action Center of Reform Judaism. A prolific writer and speaker, he has appeared on many television news and talk shows.

Introduction

I am a relatively new traveler on the Jewish-environmental path. A few years ago my stepsons, Yoni and Pesach Jeremy Stadlin, became involved with the "Teva Learning Alliance," (formerly the "Teva Learning Center,") North America's foremost Jewish Environmental Education Institute, e, a non-denominational educational service for participants from all corners of the Jewish community. Working with Jewish Day Schools, congregational religious schools, synagogues, camps and youth groups, Teva's programs provide learning workshops and seminars for some 6,000 participants annually.

Teva's web site (www.TevaLearningAlliance.org) describes its philosophy as follows:

> Thousands of years ago our ancestors lived with a keen awareness of their dependence on the natural systems that support life. Through their daily interactions with soil, water, and air, they developed a great respect for the Earth and sensed the presence of the Divine within all of Creation. Although many Jews today have lost this connection, our ancient relationship with nature is nevertheless reflected in Jewish law, in our prayers, in the celebration of our holidays, and in the core values of our tradition.
>
> The Teva Learning Alliance exists to renew the ecological wisdom inherent in Judaism. By immersing participants in the natural world and providing structured activities which [sic] sensitize them to nature's rhythms, we help them develop a more meaningful relationship with nature

and their own Jewish practices. This process also facilitates personal growth, community building, and a genuine commitment to *tikkun olam*, healing the world. All Teva programs are built on a thematic progression - from Awareness to Interconnectedness to Responsibility.

Yoni and Pesach served as counselors and educators at Teva for several years, as their commitment to the environment deepened in many ways. Their dedication to Jewish environmentalism and the sustainability of our planet grew and expanded as time passed, and their involvement branched out into their personal lives and their entire philosophy of life.

This commitment could not help but spill over into the lives of our entire family. While I was for many years devoted to recycling and healthy eating, I was not aware that these small measures were a part of a much larger philosophy of life, and I did not realize how much protecting the environment was a part of our Jewish tradition. I am indebted to Pesach and Yoni for being teachers, a great inspiration, as well as agents of change for my wife Maxine and me.

So many things in our life changed under their influence – including the way we eat, the way we drive, the way we care for our home, the way we live our lives. I cannot say enough about the amazing influence these two young thirty-somethings have impacted the lives of their parents, people somewhat habituated in their lifestyle, now in their sixties and seventies.

Yoni and his wife Vivian are, as I write, embarking on one of the most incredible, creative, innovative projects in American Jewry. They are founders and directors of the first Jewish environmental overnight summer camp in Putnam County, New York (an hour north of New York City) – Eden Village Camp. Hoping to start with some seventy to eighty campers, their first season (which began in late June, 2010) began with some 135 campers, while the second season (summer, 2011) involved over 230 participants (www.EdenVillageCamp.org). The success of the camp is astounding, reflecting the interest of parents and families in living a sustainable life in an effective educational environment. Pesach Jeremy also played an indispensable role in getting the camp off the ground, and making it the great success that it has become, through his charismatic leadership, his amazing rapport with young people, his exciting ability to provide inspiring music, and his deep commitment and experience in helping to create a better world.

The camp has already, in this early stage, achieved world-wide acclaim, with news articles appearing in many Anglo-Jewish weeklies, as well as in the prominent Israeli paper, *Haaretz*. A remarkable staff from all over the world, experts in their various fields, has been assembled by Yoni and Vivian, to launch the new camp. The encouragement and financial support of The Jim Joseph Foundation, The Foundation for Jewish Camp, and the New York UJA/Federation, were all instrumental in helping to initiate Eden Village Camp.

Our entire family could not help but be powerfully impacted by the founding of this important venture. However, this was only the climax of a lengthy period of time in which issues of environment (and its Jewish roots) and sustainability became central to our family's mode of thinking, being and acting.

As I began to explore these themes, and gradually made significant changes in our family's life (such as eating an increasingly organic diet, being more conscious of water use, transportation, "reducing, re-using and recycling," and other related areas), the idea germinated within me to write this book. While the number of "green" books on the shelves of the bookstores has been increasing in geometric proportion, and even the number of books, articles and websites on Judaism and the environment has grown greatly, I did not find one single book devoted exclusively to specific *actions* with which Jews can help "green the planet."

Thus, I set about researching the topics in this book more intensely, and putting together what can be considered a manual for Jews, Jewish families and institutions, to green their lives and communities, through a Jewish lens, and become more aware and more active in changing themselves, their awareness, their surroundings, and ultimately the planet. Non-Jews may also be interested in what the rich Jewish tradition teaches and advocates for sustainability.

Why a book on Jewish-environmental actions?

Today, our world is on an unsustainable path. While over geologic time, fossil fuels are indeed renewable, we are using this resource more quickly than those natural processes can regenerate them[1,2]. Compounding this problem is the increasing demand on non-renewable resources such as minerals.[3] Given these trends, it is not likely that our actions can be sustained indefinitely. Our children and grandchildren will have reduced access to the resources we enjoy today and will not have the opportunity to live the same high quality of life that we do, unless we change our path and identify new ways to use resources that are more sustainable.

One example is the wasteful and inefficient usage of our water resources, and the contamination of water with a host of chemicals from pollution to cosmetics. Agricultural, domestic and industrial pollution are sources for a toxic brew of heavy metals such as cadmium, lead, and mercury, as well as organic compounds such as pesticides, personal care products, and pharmaceuticals in our lakes, streams and groundwater.[4] Of special concern to the Jewish community is the increasing scarcity and decreasing quality of water in the land of Israel.[5] Additionally, availability and distribution of high quality water has always been, and continues to be, a source of political tension and violence both in the United States and around the world.[6]

Another example of this unsustainable path is the energy crisis. We currently rely primarily on non-renewable, polluting resources such as oil and coal to fuel our cars and for electricity in our homes.[7] We have built our societal structures on this unsustainable pattern, which depends on a foundation of ancient organic matter: fossil fuels such as coal, oil, and natural gas. An oil-based society is unsustainable, as available oil supplies are already starting to decline in spite of the fact that new deposits are still being identified.[8]

In addition to the impacts on air quality, health, and political security[9,10,11,12,13,14], a consensus of scientists agrees that the combustion of fossil fuels is the main driving force of global climate change.[15] These actions are also affecting the climate systems of our planet, and will cause significant consequences for future generations, such as increased storms, heat waves, and altered distribution of water resources.[16] Although public opinion still shows a dangerously low level of awareness and acceptance about climate change, the facts remain clear that the Earth's climate has warmed significantly over the last 50 years, and unless human activities change dramatically, will continue to warm throughout this century. It has been predicted by sophisticated computer models, demonstrated by carefully designed experiments, and confirmed by countless direct observations that increased concentrations of greenhouse gases in the atmosphere have profound effects on the Earth's temperature, sea level, and the abundance and distribution of precipitation.[17] (For a statement on climate change and the Jewish responsibility to help, see Figure 1.)

Not only do we have an environmental problem, we also have a justice problem, in that Americans consume significantly more of our planet's limited resources than most of the rest of the countries in the world. For example, in the US we use 68.672 barrels of oil/day per 1,000 people, while in India it is only 2.409 barrels/day per 1,000 people. Israelis consume 33.004 barrels/day per 1,000 people, just slightly above the world's average.[18]

The world must change its ways. We must change our ways. Jews have always been in the forefront of ways to change the world, improve it, repair it. (*Tikkun Olam – repairing the world* - is the Hebrew phrase that has become popular in the last few decades). Environmental sustainability is a long-term maintenance of critical resources within a context of the needs of human civilization. For example, we balance the need for specific non-renewable resources with a long-term, sustainable strategy that reduces the rate of depletion of that resource.[19]

Even if the naysayers are right, and climate change is not a pressing problem, more sustainable living will still result in positive benefits for all of us. Clean air, clean water, healthier food, tighter communities – all of these things can be benefits from an environmentally sustainable way of life. If we transition to a more renewable-based energy system, air will be cleaner, the chance of oil spills and other disasters will be reduced, pristine lands will not be threatened by drilling plans, and energy politics will no longer be at the center of global affairs. If we can make some simple yet serious changes, we still have time to reverse the trends of the past decades that have devastated our precious environment. As a human species, living more sustainably and avoiding the danger of climate change might be the finest thing we have ever done.

Yoni has taught me: "If we bite the earth, the earth bites us back." In other words, unless we treat the planet with respect, it will not be habitable in the way we have been accustomed to. If we selfishly and thoughtlessly pillage the earth our planet will be immeasurably diminished.

The Torah in the Book of Genesis tells us that God's initial design for us humans was to live in a garden of abundance. The Book of Genesis reminds us, and modern science teaches that we were created after the trees, after the animals.

The Talmud (Tractate Sanhedrin 38a) asks: Why was man created last? So in case his heart grows proud, one might say – even the gnat was created before you.

And Rashi, on "Urdu" in Chapter 1 says: Urdu:

Zacha, Rodeh, If man fulfills his mission of rulership appropriately, Rodeh – he will rule.

Im Lo Zakhah, Yarood –. And if he does not, then he will not be worthy and "yarad" he will descend.

Ora Sheinson, President and co-founder of Canfei Nesharim, teaches that this means:

> *If you are worthy – if you subdue the land while taking into consideration the effects your actions have on the future, if your actions show a stewardship of the land, then you will continue to merit rulership of the land, and you will merit Hashem's Bracha. However, if your actions show a disregard for the forces of nature, and do not account for the future consequences of today's actions, nature will overcome you and you will descend – you will be beholden to it.*

We've already cleared over 50% of our planet's forests, poached its animals and endangered species, mostly by destroying habitat.[20,21] If we want to hand over to future generations a world that has enough energy, clean air, clean water, food and other life-giving commodities, we need to act in a responsible way that will "sustain" human existence far into the future.

One can become frightened by the apparent limits to our resources. However, there are a wide range of renewable resources which are available to us if we explore "renewable" options. Renewable resources are those that regenerate themselves within a short period of time. For example, we are constantly bathed by an over-abundance of sun energy. The oceans are constantly raging with motion. Geothermal energy sources are found throughout the world. Key regions in our land- and seascape have sufficient and constant wind to tap. Biofuels from switchgrass can be grown more quickly and with less impact than ethanol.

Robert Anton Wilson, in a review of *Critical Path* by Buckminster Fuller, writes that "humanity has already achieved, technically, the total success all Utopians ever dreamed of; our problems now are entirely due to wrong thinking. We are in the tragic-comic predicament of two crazed men dying of thirst, fighting over a teaspoon of water in the middle of a rainstorm. We cannot see the rainstorm because we are hypnotized by emergency reflexes fixated on the teaspoon."[22]

Thus, the measuring rod with which we judge how we eat, how we build, how we create, how we increase, how we do just about everything, has to be this: are we sustaining life for future inhabitants of our world? Is our lifestyle a *sustainable* one, or not?

The responsibility of the Jewish People to help lead the world to a greater sense of a sustainable environment is a great one. We Jews are, according to the prophet Isaiah, "a light to the nations." What kind of light? Sustainable light!

If we can achieve this, we will be standing on the shoulders of our ancient Jewish teachers who provided wisdom that can still be used today.

While a number of books already exist on the theme of the connection between Judaism and the environment, there is not one that I am aware of that presents a complete picture of the *actions* each of us is called upon to perform in order to do our share in living a sustainable life. In this book I try to present some of these necessary actions, within a Jewish perspective, in all aspects of our lives, - at home, at work, in our synagogue and other institutions, in the community, and throughout the planet. If more people will transform their day-to-day actions in this direction, we will surely help to bring about what Jewish tradition has come to call "tikkun olam," the refining and repairing of our broken world.

Figure 1

Canfei Nesharim's Statement on Climate Change, as referenced in the introduction.

STATEMENT ON CLIMATE CHANGE
Approved by Canfei Nesharim Board of Directors, May 2009
(www.canfeinesharim.org)

Summary: What is Happening?

The Earth's climate has warmed significantly over the last 50 years, and unless human activities change dramatically, will continue to warm throughout his century.

- Fossil fuels are burned as a result of things we do every day to eat, travel, and live in our homes.
- Greenhouse gases, such as carbon dioxide, nitrous oxide and methane, are released by the burning of fossil fuels (such as oil, coal, and natural gas).
- These "greenhouse gases" remain in the atmosphere for a long time and are causing our atmosphere to become warmer.
- In addition, destruction of forests and other changes in the way we use our land have increased carbon dioxide in the atmosphere.
- The rise in atmospheric greenhouse gases parallels the rapid rise in the Earth's surface temperature during the same period of time.

It has been predicted by sophisticated computer models, demonstrated by carefully designed experiments, and confirmed by countless direct observations that increased concentrations of greenhouse gases in the atmosphere have profound effects on the Earth's temperature, sea level, and the abundance and distribution of precipitation.

Why Climate Change is Important

Global warming affects the health and welfare of humans and other living organisms on this planet. To date, the earth's surface has warmed an average of 0.7°C and according to IPCC scientists, if no corrective action takes place, carbon dioxide concentrations will double by the latter half of this century, causing a rise of from 2.0 to 5.5°C (3.6 to 10°F) in the earth's average surface temperature.[23]

While these temperature changes may seem small, it is important to understand that even these slight changes in the average surface temperature can have profound effects on the global systems that regulate our planet. One way to help put these temperature changes into perspective, one must understand what climate change means in terms of human health and welfare.[24]

While climate change might allow those in northern latitudes to enjoy warmer winters, warmer winters generally reduce the snow pack that provides the drinking water resources for a region. Snow and ice from mountain glaciers and snow pack are primary sources of water for both surface water (lakes and rivers) and ground water (the water source for wells). For example, scientists have shown that glaciers are melting at an alarming rate and may disappear within 30-50 years,[25] with serious impacts on the water availability in those regions. Additionally, in many subtropical and tropical regions, increased heat will cause drying, expanding arid conditions.

Our water resources are used for agricultural, commercial, and domestic purposes. As population numbers increase and the demands for water rise, decreasing levels of available, usable water could create severe socio-political crises and economic disruptions.

Beyond the need for water, a warmer planet can have significant impacts on agriculture. Warmer winters, earlier snowmelts and later frosts have been shown to cause greater insect infestations for agricultural crops. That is because many insects die during extended winter periods of subzero temperatures.[26] Additionally, while some species of crops grow better with increases in atmospheric carbon dioxide, the warmer, drier weather in some regions that are associated with increased carbon dioxide reduces overall productivity for many other key agricultural species.[27]

There is also a risk of warmer weather to human health. Aside from the very real risks posed by drought and heat stress (particularly to vulnerable populations such as babies and the elderly), there is also a potential for increased disease. Because changes in rainfall patterns and temperature regimens may extend the range of mosquitoes that can spread malaria, yellow and dengue fever, populations that had not experienced waterborne diseases may now be susceptible.[28]

While many of us have heard predictions of sea level rise as a consequence of climate change, we do not realize that ocean islands may be completely submerged by this increase. This reality may create millions of environmental refugees as people flood into areas in search on potable water and arable land. A recent report from the United Nations projects that the sea level could rise as much as two meters by the end of the century. Sea level rise could have devastating impacts on the hundreds of millions of people currently living in densely populated coastal regions or mega-deltas of Asia and India. The United Nations estimates that a sea level rise of only one meter will cause some areas to be uninhabitable, such as the oceanic island nation, Maldives, where about 80 percent of the 1,192 coral islets are one meter or less above sea level.[29]

The displacement of entire societies, most of whom have had no share in the causes of climate change, pose startling issues for human justice.

Scientific Evidence

Largely as a result of increased human activities, the atmospheric concentration of carbon dioxide has reached 385 parts per million, a 40% increase over levels in the pre-industrialized world[30] and the highest values known for the last 1 million years.[31] Atmospheric concentrations of carbon dioxide continue to increase at a rapid rate[32] with 75% of the increase in carbon dioxide emission occurring in the last half century.[33]

Because of the long lead-time between emissions and consequent atmospheric temperature change, the atmosphere is expected to warm another 0.6°C (1.1°F) due to emissions that have already taken place even if no additional greenhouse gases are added to the atmosphere.[34]

There is a strong scientific consensus that rising carbon dioxide and other greenhouse gases has caused temperatures to rise above pre-industrial times by about 0.7°C (1.3°F) and has brought the hottest weather on record for several years within the last decade.[35] Damage from climate change is predicted to be most severe for changes above 2°C (3.6°F) and a carbon dioxide concentration of 450 parts per million.[36] Combining these data, it is clear that the system has already experienced or is committed to some 65% of the warming we can safely afford.

Without any corrective action, carbon dioxide concentrations will double by the latter half of this century, resulting in a rise of from 2.0 to 5.5°C (3.6 to 10°F) in the earth's average surface temperature.[37] Therefore, it is highly likely that significant changes in agriculture, health, and socioeconomic parameters will occur.

Some of the carbon dioxide emitted to the atmosphere dissolves into the ocean, thereby increasing its acidity. Ocean acidification and related worldwide changes in ocean chemistry have been shown to be detrimental to marine organisms such as reef-building corals and a host of economically important fish species.[38]

Current research cannot prove with 100% certainty that existing climate change does not have a natural component along with the human industrial influence. However, predictions climate scientists made 20 years ago as to the Earth's response to increasing atmospheric carbon dioxide levels are entirely consistent with the actual changes that have occurred.[39] This powerful finding is matched by retrospective computer modeling that has demonstrated that the single most accurate predictor of current changes in land and ocean temperatures is the human-driven emissions of carbon dioxide.[40]

While there are uncertainties in the science of climate change, they are not about whether the Earth has already warmed or whether human emissions of greenhouse gases are responsible for a large part of this warming. Rather, they concern the magnitude of future changes if civilization does not take corrective action.

What Can Be Done to Help

Given that human activities are a critical cause of global climate change, Canfei Nesharim recognizes that people can also become motivated to act to reduce or even reverse the rate at which these changes are occurring. Specifically, as an organization we appeal to the full range of societal stakeholders to promote policies and practices that reduce carbon footprints at the individual, corporate, national, and international levels.

This appeal is consistent with the Biblical (d'araiso) injunction, *Bal Tashchit*—from the book of Deuteronomy (Chap. 20: 19-20). This injunction not to waste or destroy trees in times of war has been interpreted by the rabbinic authorities of the Talmud and later generations to encompass a larger proscription against wasteful destruction of resources.

"When you lay siege and battle against a city for a long time in order to capture it, you must not destroy its trees, wielding an ax against them. You may eat of them, but you must not cut them down – for is a tree of the field a man, that it should be besieged by you? Only the trees which you know are not trees for food, you shall destroy and cut them down, and you shall build siegeworks against the city that makes war with you until it is subdued."

Over the centuries, the instruction not to waste became extended well beyond fruit trees to become a general philosophy of prudence and protection, consistent with our stewardship over the earth. We are obliged not to use more than what we need, not to destroy things needlessly, not to use something of greater value when something of lesser value will suffice and not to use something in a way it was not meant to be used (which increases the likelihood the item will be broken or destroyed). This injunction points us in the direction of proper, sustainable use of natural resources, and certainly is consistent with preventing significant damage to natural systems upon which life depends, such as our climate.

Our actions are causing destructive harm to the resources on our planet, upon which all life depends. Our Jewish tradition obligates us to take action.

Acting locally to limit our energy use and reduce our environmental impact can change a mindset of disbelief and inaction that, in turn, will move corporations and governments to greater action. Examples of these small-scale actions include but are not limited to:

- If you can walk to *shul* [synagogue] on Shabbat, walk there for daily *minyanim*
- [prayer services] whenever it is convenient.
- Get an energy audit for your congregation and, within the ability of your community to do so, implement an energy savings plan for the building.
- Consider the energy used in simple activities such as water usage in your home and synagogue; lighting, heating and cooling buildings; and growing, transporting, and buying food.
- Buy local to reduce fuel requirements in merchandise transportation both for your home and *shul*.
- Even more effective than buying local is shifting to less than one day per week's worth of calories from red meat and dairy products to chicken, fish, eggs, or a vegetable-based diet achieves more greenhouse reduction than buying all locally sourced food.[41]
- "Green" your home and shul as best as you can in terms of its physical structure, e.g., where appropriate, install a green roof or solar panels and compact fluorescent bulbs (taking care to buy those with lower mercury).
- Use carpools, bicycle, or take advantage of public transportation to get to work. When buying a new car, consider the models with the highest gas efficiency.
- Minimize home and synagogue energy usages, e.g., if you are not in a room, then turn the light off; use setback thermostats to control air temperatures; and insulate your house.
- Work with organizations that encourage greater tree plantings or roof gardens in urban areas, especially pollution-tolerant species. This will help reduce the urban heat island effect, which is associated with increased heat-related illnesses and mortalities.[42]

Section A: The Basics

The Jewish environmental path begins with three key steps.

We begin by recognizing the challenges we face globally, and see how our personal impact is a part of that global challenge. We can do this by assessing our own environmental impact in a global context.

We must take ownership of our own place, by getting to know our own local environment. The local environment includes a specific set of challenges. What are its native plants, and what invasive plants have arrived and are taking over? What green places are threatened and need protection? In addition, localities include unique resources, including local environmental organizations, supportive politicians, and waste management services. These resources can be of great assistance for the environmentally concerned.

In addition to the "think global, act local," we'd like to add a third premise. As Jews, we are heirs to a rich and wise tradition which can inform and enrich our environmental awareness and action. Protecting the environment is not only a practical matter; it is also a moral and even a spiritual one. Learning and acting on Jewish teachings on the environment can deepen our environmental understanding and help us continue our commitment even when it is difficult. It is also key to maintaining hope in challenging times.

As you begin your environmental path, you have a wide range of resources available to you in each of these areas. Ready? Let's begin!

Number One

Think Global: Measure and Reduce Your Ecological Footprint

> *You can't fool Mother Nature. She knows when we're just messing around. Mother Nature operates by her own iron laws. And if we violate them, there is no lobby or big donor to get us off the hook. No, what's gone will be gone. What's ruined will be ruined. What's extinct will be extinct —— and later, when we're finally ready to stop messing around, it will be too late.*
> Thomas Friedman, *The New York Times*, May 5, 2010

"In all your ways, acknowledge God." (Proverbs 3:6). As Jews every step we take should reflect our deep-rooted Jewish values. Just as we must act ethically toward our fellow humans, so must we act responsibly toward our planet, the gift God has given us to inhabit while we walk this earth. In the environmental movement it is especially important to remember that whatever we do has a consequence. We call thi our "footprint."

Your environmental "footprint" is the area of land and ocean required to support your consumption of food, goods, services, housing, and energy and assimilate your wastes. It is the best way to measure your impact on the environment and compare it with others around the world. It's a good idea to measure your environmental footprint before you start making environmental changes. One way of thinking of environmental change is as a "diet" – before you get started, you'll want to know what you weigh! Taking an environmental footprint quiz will help you see where you are and learn the best opportunities to reduce your impact.

Recommended Action:

Calculate your ecological footprint. Here are two websites to help you:

http://www.myfootprint.org/

http://www.footprintnetwork.org/en/index.php/GFN/page/calculators/

An Ecological Footprint tool is a much more comprehensive measure of your impact on the environment than some "carbon footprint calculators" you may have seen. For that reason, we recommend that you use an Ecological Footprint quiz to get started on your environmental journey.

The Global Footprint Network[43] aims to measure how many resources we have, how much is being used, and who is using it. You can measure your Ecological Footprint with their online tool at http://www.footprintnetwork.org/en/index.php/GFN/page/calculators/.

Another commonly used website was developed by the Center for Sustainable Economy (CSE). In this tool - http://www.sustainable-economy.org/ - your ecological footprint is expressed in "global hectares" (gha) or "global acres" (ga), which are standardized units that take into account the differences in biological productivity of various ecosystems impacted by your consumption activities. The footprint is broken down into four consumption categories:
- carbon (home energy use and transportation),
- food,
- housing, and
- goods and services.

Your footprint is also broken down into four ecosystem types or biomes:
- cropland,
- pastureland,
- forestland, and
- marine fisheries.

The website also shows global averages for these elements, and compares your ecological footprint to a "sustainable footprint," which means, it compares how you live to the Earth's biological carrying capacity (global hectares or global acres divided by its population). According to the Center for Sustainable Economy, the current estimate is that a "sustainable footprint" means that each person on earth can sustainably use 15.71 global hectares or 43 global acres.[44]

Whichever website you use, after you take the quiz, the website will recommend specific actions to help you reduce your impact. Some possible actions are in the areas of energy savings, waste, purchases, food, home choices, and lifestyle. In each of our actions in these areas, we reflect our Jewish values. We will discuss these actions in more detail throughout the rest of this book.

Put it into Action!

My Ecological Footprint Results:
_____
_____
_____
_____

Goals for future action:
_____
_____
_____
_____

Number Two

Act Local: Get to Know Your Local Environment

"Television tells us we have everything in common. But we don't.... Each piece of land the world over is different – the climate and the topography and the vegetation all combine to mean that the field over there is particularly vulnerable to erosion or that the deer need that bit of stream or that the groundwater's shallow here and easily depleted. These kinds of lessons – and the affection needed to implement them – can be learned only by long observation."
Bill McKibben, *The Age of Missing Information*, pp. 40-41

Our technological age often makes it seem that instead of living in a place, we live inside the computer or the television. However, we do live in a place – a unique part of our planet, inhabited by a multitude of other species, and with specific challenges and opportunities. If we think of the place in which we live as a gift from God, with a sense of awe and gratitude for the blessings of our local environment, our attitude and behavior toward it will have specific consequences, reflecting our Jewish value system.

> **Recommended Action: Get to know your place.**
> - Explore natural spaces in your area.
> - Seek out local resources to support you in your environmental efforts.
> - Find out what challenges exist in your area and what you can do about them.

While we may not have the time or ability for long observation of the land, we can still become aware of the specific natural features of our local environment.

For example, ponder the following questions. What green spaces are near my home, and are they well tended? Where is the water? Where is the watershed boundary? What animals live in the woods? Which plants do best in the garden, and which will take over other plants? These types of questions can help us get to know our environment, and by getting to know our own local environment, we take ownership of our place.

If you live in a city, find out the history of the little streams or giant trees in your area; those are the remnants of an older ecological system which is still operating at some level. Many cities also have parks, sometimes divided by highways or paved running paths, but still teeming with local wildlife.[45] If you don't have any local parks, you may want to look for opportunities for urban gardens, or schedule a trip to the nearest green area.

If you live in a place with easy access to nature, make some time to see it. Schedule a local hike or simply spend a few hours in a park with your family. Explore the natural spaces. Take note of the local wildlife. Why is that soil eroding? Where did that creek come from and where is it going? Why does the water end there?

Depending on where you live, there may be resources available to help you. Local nature centers and even zoos can offer a wealth of environmental information about your area. Look for local environmental chapters of national environmental organizations.

Find out what local farms are in your area, and where they sell their produce. Buying local food is a great way to find out what foods are in season and to feel more connected to your place. If the local food is organic or grown using "integrated pest management" techniques which reduce pesticides, all the better. (For more information, see #25: Buy Local.)

Learn about the environmental records of your local and national representatives, and let them know that you are concerned about protecting local spaces and promoting environmental action.

Investigate the environmental policies of your county or city. What items are accepted for recycling? Do they compost yard waste? Do they compost food? Some localities even provide compost bins and will teach you how to use them. Do they have specific opportunities for the recycling and proper disposal of electronic waste? As you get to know the resources available, share them with others in your community so they can take advantage, as well.

Your local energy company may also be an ally, providing energy audits or other technological options to help save energy. You can also find out if they offer renewable energy. (For more information, see #6: Green Energy.)

Another resource in your community may be its public transportation system, which can enable you to reduce your driving. For more details on public transportation in your community, visit http://www.publictransportation.org/.

To find local, fresh, sustainable food in your area, check out this web site: www.eatwellguide.org. All you have to do is enter your zip code and it will help you find everything you need.

A deeper sense of place will help you to protect the resources that are most important in your locality, the place where you have the most ability to make a difference. But the changes you make will also have a global impact. All of our actions should manifest our attitude of gratitude towards our Maker, Who provided us with a place to live.

Put it into Action!

Getting to Know My Place

Unique features of my local environment:

_____
_____
_____
_____

Local resources (environmental groups, nature centers, waste services, energy options):

_____
_____
_____
_____

Local challenges (invasive species, water shortages, flooding, etc.):

_____
_____
_____
_____
_____

Number Three

What's a Green Jew? Learn What the Torah Says About Protecting the Environment

"When God created the first human beings, God led them around the garden of Eden and said: "Look at my works! See how beautiful they are—how excellent! For your sake I created them all. See to it that you do not spoil and destroy My world; for if you do, there will be no one else to repair it." (*Midrash Kohelet Rabbah*, 7:28 on Ecclesiastes 7:13)

Recommended Actions:

Learn what Jewish tradition says about protecting the environment.

- Review the source sheets in Figures 2-4.
- When you read Jewish books and when you hear the Torah read, consider what these lessons can teach us about how to treat our environment.
- Consider signing up for regular emails from a Jewish environmental organization.

Protecting the environment is not only a practical matter; it is also a moral and even a spiritual one. As Jews, we are heirs to a rich and wise tradition that can inform and enrich our environmental awareness and action.

In 1990, scientists wrote an "Open Letter to the Religious Community"[46] calling upon religious bodies to enter the environmental movement. "Problems of such magnitude," world-renowned scientists such as Carl Sagan attested, "and solutions demanding so broad a perspective, must be recognized from the outset as having a religious as well as a scientific dimension." The scientists go on to urgently appeal to the religious community to commit boldly to preserve the environment.

The scientists recognized that technological and scientific answers to environmental problems are not sufficient. Religion can offer a moral and spiritual grounding, which can help people change not only what they do but also, more importantly, how they think. Ultimately, changes in thinking are necessary to address the environmental crisis.

Today's environmental movement has sometimes been accused of being too negative, issuing "doomsday" predictions without providing enough inspiration or a clear path toward redemption. Jewish faith helps us maintain hope in times of great challenge, empowering us to act.

Numerous Jewish environmental organizations have created a wealth of resources to enrich our environmental action with a Jewish context. For example, you can learn what the Torah says about the environment each week by signing up for Canfei Nesharim's weekly emails at www.canfeinesharim.org. A list of organizations is provided in Number Twenty Nine: Get to Know the Jewish Environmental Movement.

Put it into Action!

Being a Green Jew

After you've read the source sheet in Figures 2-4, take a few moments to reflect on what you learned.

I didn't know that:

_____

_____

_____

I am inspired by:

_____

_____

I'd like to learn more about:

_____

_____

_____

Figure 2

Ten Teachings On Judaism and Environmentalism
by Rabbi Lawrence Troster
Director, GreenFaith Fellowship
Program (www.GreenFaith.org)

1. *God created the universe.*

This is the most fundamental concept of Judaism. Its implications are that only God has absolute ownership over Creation (Gen. 1-2, Psalms 24:1, I Chron. 29:10-16). Thus Judaism's worldview is theocentric not anthropocentric. The environmental implications are that humans must realize that they do not have unrestricted freedom to misuse Creation, as it does not belong to them. Everything we own, everything we use ultimately belongs to God. Even our own selves belong to God. As a prayer in the High Holiday liturgy proclaims, "The soul is Yours and the body is your handiwork." As we are "sojourners with You, mere transients like our ancestors; our days on earth are like a shadow..." (I Chronicles 29:15), we must always consider our use of Creation with a view to the larger good in both time (responsibility to future generations) and space (others on this world). We must also think beyond our own species to that of all Creation.

2. *God's Creation is good.*

In Genesis 1: 31 when God found all of Creation, "very good," this means several things. First of all it means that Creation is sufficient, structured and ordered (the rabbis called it *Seder Bereishit*, the Order of Creation). It is also harmonious. It exists to serve God (Psalm 148). This order reflects God's wisdom (Psalms 104:24), which is beyond human understanding (Psalms 92:6-7, Job 38-39). All of God's creations are consequently part of the Order of Creation and all are subject to its nature (Psalm 148). Humans are also part of the Order, which can be said to be a community of worshipers.

3. *Human beings are created in the image of God.*

Human beings have a special place and role in the Order of Creation. Of all God's creations, only human beings have the power to disrupt Creation. This power, which gives them a kind of control over Creation, comes from special characteristics that no other creature possesses (Psalm 8). This idea is expressed in the concept that humans were created in the image of God (*tzelem Elohim*). In its original sense, *tzelem Elohim*, means that humans were put on the earth to act as God's agents and to actualize God's presence in Creation.

This also has ethical implications that stem from the fact that human beings have certain intrinsic dignities: infinite value, equality and uniqueness. It also means that human beings possess God-like capacities: power, consciousness, relationship, will, freedom and life. Human beings are supposed to exercise their power, consciousness and free will to be wise stewards of Creation. They should help to maintain the Order of Creation even while they are allowed to use it for their own benefit within certain limits established by God (Genesis 2:14). This balance applies to both human society as well to the natural world. Since the time of the expulsion from Garden of Eden, Creation has tended to be out of balance because of the human impulse towards inequality resulting from the misuse of its powers for selfish ends. The earth is morally sensitive to human misdeeds (Genesis 4, Leviticus 18:27-30).

4. *Humanity should view their place in Creation with love and awe.*

It may be said that there are two books of God's revelation to humanity: The Torah and Creation itself. The book of Creation can help us to perceive ourselves as "living breathing beings connected to the rhythms of the earth, the biogeochemical cycles, the grand and complex diversity of ecological systems." (Mitchell Thomashow, *Ecological Identity*)

This knowledge is gained both through an understanding of Creation through scientific knowledge. In Judaism, this can be understood as the fulfillment of the commandments to love and to fear God (Deuteronomy 6:5,13). Rambam (Moses Maimonides, 1135-1204) interpreted these commandments in the following way:

"When a person observes God's works and God's great and marvelous creatures, and they see from them God wisdom that is without estimate or end, immediately they will love God, praise God and long with a great desire to know God's Great Name...And when a person thinks about these things they draw back and are afraid and realizes that they are small, lowly and obscure, endowed with slight and slender intelligence, standing in the presence of God who is perfect in knowledge." (*Mishneh Torah, Sefer Madah, Hikhot Yesodei Ha-Torah* 2:1-2)

Thus, when we study Creation with all the tools of modern science, we are filled with love and a sense of connection to a greater order of things. We feel a sense of wonder but also a sense of awe and humility as we perceive how small we are in the universe as well as within the history of evolution. Love and humility should then invoke in us a sense of reverence for Creation and modesty in our desire to use it. We should, according to Rabbi Abraham Joshua Heschel, see the world as God-centered, not human-centered. By putting God at the center of life, we see the sacred in everything and the natural world becomes a source of wonder and not only a resource for our use and abuse.

5. *The Sabbath and prayer help us to achieve this state of mind.*

The Sabbath is a way to begin to engender this sense of love and humility before Creation. It is also is a way to living a sustainable life. For one day out of seven, we limit our use of resources. We walk to attend synagogue and drive only when walking is not possible. We do not cook and we do not shop. We can use the day for relaxation, contemplation and to ask ourselves: what is the real purpose of human life? Are we here on earth only to get and to spend? As Rabbi Ismar Schorsch has written: "To rest is to acknowledge our limitations. Willful inactivity is a statement of subservience to a power greater than our own." (*To Till and to Tend*, page 20)

Prayer also helps us to recognize that everything we are, everything we have and everything we use ultimately comes from God (*Babylonian Talmud*, B'rakhot 35a). When we say a blessing, we create a moment or holiness, a sacred pause. Prayer also creates an awareness of the sacred by taking us out of ourselves and our artificial environments and allowing us to truly encounter natural phenomenon. Prayer creates a loss of control which allows us to "see the world in the mirror of the holy." (Heschel) We are then able to see the world as an object of divine concern and we can then place ourselves beyond self and more deeply within Creation.

6. *The Torah prohibits the wasteful consumption of anything.*

In Judaism, the halakhah (Jewish law) prohibits wasteful consumption. When we waste resources, we are violating the mitzvah (commandment) of *Bal Tashkhit* ("Do not destroy"). It is based on Deuteronomy 20:19-20:

"When in your war against a city you have to besiege it a long time in order to capture it, you must not destroy its trees, wielding the ax against them. You may eat of them, but you must not cut them down. Are trees of the field human to withdraw before you into the besieged city? Only trees that you know do no yield food may be destroyed; you may cut them down for constructing siegeworks against the city that is waging war on you, until it has been reduced."

This law was expanded in later Jewish legal sources to include the prohibition of the wanton destruction of household goods, clothes, buildings, springs, food or the wasteful consumption of anything (see Rambam, *Mishneh Torah*, *Laws of Kings and Wars* 6:8, 10; Samson Raphael Hirsch, *Horeb*, 279-80). The underlying idea of this law is the recognition that everything we own belongs to God. When we consume in a wasteful manner, we damage Creation and violate our mandate to use Creation only for our legitimate benefit. Modesty in consumption is a value that Jews have held for centuries. For example one is not supposed to be excessive in eating and drinking or in the kind of clothes that one wears (Rambam, *Mishneh Torah*, *Laws of Discernment*, chapter 5). Jews are obligated to consider carefully our real needs whenever we purchase anything. We are obligated when we have a *simchah* (a celebration) to consider whether we need to have elaborate meals and wasteful decorations. We are obligated to consider our energy use and the sources from which it comes.

7. *The Torah gives an obligation to save human life.*

The Jewish tradition mandates an obligation to save and preserve life (called in Jewish legal sources: *pikuach nefesh*) based on an interpretation of Leviticus 18:5, "You shall keep My laws and My rules, by the pursuit of which man shall live: I am the Lord (See *Babylonian Talmud* Sanhedrin 74a)." Jewish law forbids us from knowingly harming ourselves (Leviticus 19:28). There are also numerous sources mandated the proper disposal of waste is properly and that noxious products from industrial production must be kept far from human habitation (see for example, Deuteronomy 23:13-15, *Mishnah Bava Batra* 2:9) In the Jewish tradition, the public good overrides individual desires. While there are many useful and even lifesaving technologies that come from modern chemicals and materials, we have an obligation to be cautious in their use. *Pikuach nefesh* demands that we consider the impact of our use of chemicals and other materials, not only in the short term but also in the long term. For the Jewish tradition, the Precautionary Principle can be seen as a modern form of the warning not to tamper too much with the boundaries of Creation.

8. *The Torah prohibits the extinction of species and causing undo pain to non-human creatures.*

Our ancestors could not have anticipated the loss of biodiversity that the modern world has produced; from their perspective, there was no natural extinction rate of species. God, they believed, had created all species at one time and there could be no new creatures. Only humans could cause extinction and bring about the loss of one of the members of the Creation choir. In the Torah there is a law that says:

"If along the road, you chance upon a bird's nest, in any tree or on the ground, with fledglings or eggs and the mother sitting over the fledglings or on the eggs, do not take the mother with her young. Let the mother go, and take only the young, in order that you may fare well and have a long life." (Deuteronomy 22:6-7)

Ramban (Moses ben Nahman, Nahmanides, 1194-1270) in his commentary to the Torah wrote:

> "This also is an explanatory commandment of the prohibition *you shall not kill it {the mother} and its young both in one day* (Leviticus 22:28). The reason for both [commandments] is that we should not have a cruel heart and not be compassionate, or it may be that Scripture does not permit us to destroy a species altogether, although it permits slaughter [for food] within that group. Now the person who kills the mother and the young in one day or takes them when they are free to fly, [it is regarded] as though they have destroyed that species."

> It is evident from the first chapter of Genesis and other biblical texts (Psalms 104, 148, and Job 38-41) that God takes care of, and takes pleasure in, the variety of life that makes up Creation. And although we might regard a species as unimportant or bothersome to human beings, God does not regard them so. The rabbis understood that we do not know God's purpose for every creature and that we should not regard any of them as superfluous. "Our Rabbis said: Even those things that you may regard as completely superfluous to Creation – such as fleas, gnats and flies—even they were included in Creation; and God's purpose is carried through everything—even through a snake, a scorpion, a gnat, a frog." (*Midrash Bereishit Rabbah* 10:7) In environmental terms, every species has an inherent value beyond its instrumental or useful value to human beings. Related to this idea is the concept of *Tzar Baalei Chayyim*, the prohibition of hurting animals without good purpose (based on Deut. 22:6, 22:10, 25:4, Numbers 22:32, Exodus 20:8-10, Lev. 22:27-8). These concepts bring to our relationships with the non-human world limits and controls over our power and greed.

9. *Environmental Justice is a Jewish value.*

> The Torah has numerous laws that attempt to redress the power and economic imbalances in human society and Creation. Examples are the Sabbatical year (Exodus 23:11, Leviticus 25:2-5, Deuteronomy 15:1-4) and the Jubilee (Leviticus 25:8-24) There is a whole program in the Torah for creating a balanced distribution of resources across

society (Exodus 22:24-26, Leviticus 25:36-37, Deuteronomy 23:20-1, 24:6,10-13,17). This is an expression of the concept of *Tzedek*, which means righteousness, justice and equity. It is the value, which tries to correct the imbalances, which humans create in society and in the natural world. In the modern world globalization has strived to achieve the free movement of people, information, money, goods and services but it can also create major disruptions in local cultures and environments. While globalization has created great wealth for millions of people, many millions more have been bypassed by its benefits and has had in some cases a negative impact upon the environment and human rights. The Jewish concept of *Tzedek* demands that we create a worldwide economy that is sustainable and that is equitable in the distribution of wealth and resources.

10. *Tikkun Olam: The perfection/fixing of the world is in our hands.*

There is a midrash (Rabbinic commentary on the Bible) which Jewish environmentalists are fond of quoting:

> "When God created the first human beings, God led them around the garden of Eden and said: "Look at my works! See how beautiful they are—how excellent! For your sake I created them all. See to it that you do not spoil and destroy My world; for if you do, there will be no one else to repair it." (*Midrash Kohelet Rabbah*, 7:28 on Ecclesiastes 7:13)

In the Jewish liturgy there is a prayer called *Aleinu* in which we ask that the world be soon perfected under the sovereignty of God (*le-takein olam be-malkhut Shaddai*). *Tikkun 'olam*, the perfecting or the repairing of the world, has become a major theme in modern Jewish social justice theology. It is usually expressed as an activity, which must be done by humans in partnership with God. It is an important concept in light of the task ahead in environmentalism. In our ignorance and our greed, we have damaged the world and silenced many of the voices of the choir of Creation. Now we must fix it. There is no one else to repair it but us.

Figure 3

Jewish Texts as Resources on Global Climate Change
Compiled by Rabbi Fred Scherlinder Dobb,
Adat Shalom Reconstructionist Congregation,
Bethesda MD

1. Judaism has always recognized that **the stakes are high**, and we can't afford to make lasting mistakes:

    "God led Adam around all the trees of the Garden of Eden. And God said to Adam: 'See My works, how good and praiseworthy they are?! And all that I have created, I made for you. [But,] be mindful then that you do not spoil and destroy My world - for if you spoil it, there is no one after you to repair it.'" (*Midrash Qohelet Rabbah* 7:13; ca. 8th Century C.E.)

2. Remember **whose Earth** we're on in the first place, and what we're supposed to be doing with it:

    "The Earth is God's, and the fullness thereof; the settled land, and its inhabitants." (Psalms 24:1)

    "The land shall not be sold forever; for the land is Mine; you are strangers and sojourners with me." (Leviticus 25:23)

    "God placed the human in the Garden of Eden, *l'ovdah* (to serve/ till) *u'l'shomrah* (and to guard/tend) it." (Gen. 2:15)

3. **Conservation**: Wasting anything is a shame (especially when it's so easy to use less electricity or get better mileage or...)

    Bal Tashkhit: "When you besiege a city... do not destroy (*lo tashchit*) any of its trees... you may eat of them, but not cut them down." (Deuteronomy 20:19)

Rav Zutra said: "Whoever covers an oil lamp, or uncovers a naphtha lamp, transgresses the law of *bal tashkhit*." (Talmud, Shabbat 67b. These actions make fuel burn inefficiently. See energy conservation & emissions standards!)

"Righteous people ... do not waste in this world even a mustard seed. They become sorrowful with every wasteful and destructive act that they see, and if they can, they use all their strength to save everything possible from destruction. But the wicked ... rejoice in the destruction of the world, just as they destroy themselves." (*Sefer HaHinukh* 529; 13th Century)

4. **Justice / Equality:** We in the US are under 5% of the world's population, yet cause at least a fourth of all greenhouse gases. And who will rising sea levels and other effects of climate change harm most? Poor people in developing nations....

    "*Tzedek tzedek tirdof* -- Justice, justice, you shall pursue, in order that you may live... " (Deuteronomy 16:20)

    "God loves righteousness and justice; the Earth is full of God's loving-kindness." (Psalms 33:5)

    "Do not stand idly by the blood of your neighbor ... Love your neighbor as yourself." (Leviticus 19:16, 19:18)

5. **Preserving Life:** With climate change we will likely cause the spread of new diseases, longer heat waves, more intense hurricanes, worsening agricultural losses, and potentially the social instability that accompanies these trends.

    "One is forbidden from gaining a livelihood at the expense of another's health." (Rabbi Isaac ben Sheshet, Resp. 196, 14th Century)

    "Shabbat, like all the *mitzvot*/commandments, is pushed aside by danger to human life." (Rambam, Mishneh Torah, Zmanim 2:1)

6. Saving **Endangered Species**: Everything's part of the plan, yet global warming moves too fast (hundreds of times more abruptly than 'natural' change) for most of Creation to adapt, threatening many species and whole ecosystems.

    "Even those creatures you deem superfluous in the world – like flies, fleas, and gnats -- nevertheless have their allotted task in the scheme of Creation (*seder beresheet*)." (*Midrash Exodus Rabbah* 10:1)

"It should not be believed that all beings exist for the sake of humanity's existence ... [rather,] all the other beings, too, have been intended for their own sakes... " (Rambam / Maimonides, *Guide of the Perplexed* III:13; 12th Century, Egypt)

7. The **Precautionary Principle**: Even if we accept (despite near unanimity among independent scientists) "debate" about the reality of climate change, Judaism teaches us to take serious precautionary measures, before all the data are amassed:
"When you build a new house, you shall make a parapet for your roof, so that you do not bring bloodguilt on your house if anyone should fall from it." (Deuteronomy 22:8)
"Similarly with all potentially dangerous objects. Remove them far from yourselves and from the way of the community." (Maimonides, MT *Hilchot De'ot*, 12th Century)
"A burning coal/object left in a place where the public can be injured by it - one is allowed to extinguish it [even on Shabbat], whether it's of metal or of wood." (Yosef Caro in *Shulhan Arukh, Oreh Hayim* 334:27; 16th Century Tzfat)
"A sick person in danger - we attend to all their needs on Shabbat, at the advice of skilled local healer. If there is a doubt whether or not we need to violate the Shabbat for them - or if one doctor says to, ... but another doctor says there's no need - we violate the Shabbat for them, since [even] doubtful danger to human life pushes aside the Shabbat." (Rambam, MT *Zmanim* 2:1, continuing the quote above at #5. Replace "doctor" with "scientist," and "Shabbat" with "corporate profits"?!)
"... We don't need an expert [to save a life by violating other laws like Shabbat], since ... [even] doubtful danger to human life [makes the law] lenient. And it's forbidden to delay the thing [treatment]... " (*Tur*, 14th Century Spain, OH 328 - to which Caro adds, "the one who rushes to do so, look, this is praiseworthy! But the one who [stops to] ask, look, this is a murderer.")

8. In **Conclusion**: "See, I have set before you this day life and death, blessing and curse - and [you should] choose life, in order that you and your children may live."
(Deuteronomy 30:19)

Figure 4

Pillars of Jewish Environmental Awareness
By Dr. Gabe Goldman
Director of Experiential and Environmental Education,
Brandeis-Bardin Institute
primskills@yahoo.com

Thousands of years before the word "environmentalism" was coined, Jewish tradition paid attention to taking care of the earth and treating animals with compassion. Thousands of years before the first landfill appeared, the Torah was teaching Jews not to waste anything. Earth care has always been important in Judaism, partly because nature is so integral to Jewish life. Nature imagery fills our prayers and enhances many of our holiday celebrations. Our rabbis of old recognized that they could better understand the Creator by observing creation. To this end, they spent many hours in the outdoors— watching the sun rise and set, noting the phases of the moon, delighting in the changing of the seasons.

The pillars of Jewish environmental awareness are rooted in Judaism's cosmological beliefs (beliefs about how the world was created) and expressed as halakhah (Jewish law) in the Torah and Talmud, our "written" and "oral" traditions. In my view there are eight fundamental beliefs that govern Jewish environmental awareness. These are:

1. *The Belief in the Oneness of God*
The most basic belief of Judaism is the belief in one God. This belief is made the central prayer of the Jewish worship service, the Shema: *Listen Israel, the Lord is our God, the Lord is One*. The first of the Ten Commandments tells us that there is only one true God and that we are not to worship false gods. So, too, Judaism tells us that this Oneness is evident in the inter-connectedness of the natural world. Science describes the interconnectedness as "ecology."

2. *God is the "Owner" of the World*
Judaism views God as the rightful owner of the world. Leviticus (25:23) states this explicitly, *The Land shall not be sold for eternity; for the land is mine and you are but strangers journeying with Me."* We are also the caretakers of this world, a responsibility assigned to all humans the moment God placed Adam in the Garden of Eden to "Work and protect" (*avdah u-shomrah*) it. We have a right to use any of its resources but this right is tempered by our responsibility to protect these resources for use by all future generations.

3. *God Created the World with Intent and Purpose*
Jewish tradition tells us that God created the universe with purpose. Nothing was created by "accident" or without a reason. A traditional Jewish story makes this point beautifully. *Midrash Bereshit Rabbah (10:7) - Even though you may think them superfluous in this world, creatures such as flies, bugs and gnats have their allotted task in the scheme of creation, as it says, "And God saw everything that God had made, and behold, it was very good" (Genesis 1:31).*

4. *Earth Stewardship is the Responsibility of the Individual*
Like most of the Torah's commandments, taking care of the earth is made the responsibility of the individual. Earth stewardship is not made the responsibility of political parties, environmental movements or religious organizations. According to a traditional Jewish story, this point was made dramatically clear by God to the first man and woman. *Midrash Ecclesiastes Rabbah (7:28) - When God created Adam, God led him around all of the trees in the Garden of Eden. God told him, "See how beautiful and praiseworthy are all of my works. Everything I have created has been created for your sake. Think of this and do not corrupt the world; for if you corrupt it, there will be no one to set it right after you."*

5. *Baal Tashkhit* -- **Prohibitions against Waste**
Called *bal tashkhit* in Hebrew, this commandment is the basis of the prohibition against wasting or destroying anything needlessly. The prohibition is found in Deuteronomy (20:19-20), *When in your war against a city you have to besiege it for a long time in order to capture it, you must not destroy (bal tashkhit) its fruit trees. . . You may eat of them but you must not destroy the fruit trees.* Later Jewish thinkers explained that *bal tashkhit* applies to every person all of the time, encompassing the prohibitions against using more of

something than is necessary, using something in a way it is not intended to be used and using something of greater value when something of lesser value could be used.

6. *Tzaar Ba'alei Hayyim* -- Prohibition against Causing Animals Unnecessary Pain

Called *tzaar ba'alei hayyim* in Hebrew, this prohibition tells us not to cause animals any unnecessary physical or emotional pain. So important is this prohibition that it appears in several places throughout the Torah, one such being Deuteronomy (5:14) - *If, on your way, you happen upon a bird's nest in a tree or on the ground, with baby birds or eggs in it, do not take the mother with her young. Drive away the mother and take only the young. This way you will live a long life.*

7. *Shabbat and Sabbatical Years* -- Land Rest and Renewal

Land rest and renewal are concepts that appear in the Torah in conjunction with instructions on how to care of the Land of Israel. The earth is to rest once a week on Shabbat and every seventh year, called the *Sh'mitah* or Sabbatical year. The instructions regarding the Sabbatical year are found in Leviticus (25:3-4) - *Six years you may sow your field and six years you may prune your vineyard and gather in the field. But in the seventh year the land shall have a sabbath of complete rest, a sabbath of the Lord: you shall not sow your field or prune your vineyards.*

8. *Earth Stewardship is a Personal Commitment*

Observing the mitzvot of earth stewardship is a personal commitment. It is not a commitment that depends on a grass-roots party or a supportive political system. Like almost all of the mitzvot in the Torah, it is the responsibility of the individual to perform regardless of what others are doing. Ultimately, Jewish tradition believes that the individual who performs the mitzvot with joy will have a positive effect on those who do not. Ultimately, this is the lesson we learn from Abraham's role in bringing monotheism to the world.

Section B: So Many Ways to Save Energy

"Remember that it is God Who gives you the power..." Deuteronomy 8:18

"They who trust in God shall renew their energy" Isaiah 40:31

If you're like most people, you get your energy from the grid. According to the United States Environmental Protection Agency, electricity generation from fossil fuel-fired power plants is responsible for 67 percent of the nation's sulfur dioxide emissions, 23 percent of nitrogen oxide emissions, and 40 percent of man-made carbon dioxide emissions. These emissions can lead to smog, acid rain, and haze. In addition, the greenhouse gas emissions from power plants are one of the primary causes of climate change.[47]

Coal is one of the most common energy sources for power plants, and the dirtiest. Coal pollutes when it is mined, transported to the power plant, stored, and burned. There are about 600 U.S. coal plants. A typical (500 megawatt) coal plant burns 1.4 million tons of coal each year.[48] In an average year, a typical coal plant generates:
- 3,700,000 tons of carbon dioxide (CO_2), the primary human cause of global warming--as much carbon dioxide as cutting down 161 million trees.
- 10,000 tons of sulfur dioxide (SO_2), which causes acid rain that damages forests, lakes, and buildings, and forms small airborne

- particles that can penetrate deep into lungs.
- 500 tons of small airborne particles, which can cause chronic bronchitis, aggravated asthma, and premature death, as well as haze obstructing visibility.
- 10,200 tons of nitrogen oxide (NO_x), as much as would be emitted by half a million late-model cars. NO_x leads to formation of ozone (smog) which inflames the lungs, burning through lung tissue making people more susceptible to respiratory illness.
- 720 tons of carbon monoxide (CO), which causes headaches and place additional stress on people with heart disease.
- 220 tons of hydrocarbons, volatile organic compounds (VOC), which form ozone.
- 170 pounds of mercury, where just 1/70th of a teaspoon deposited on a 25-acre lake can make the fish unsafe to eat.
- 225 pounds of arsenic, which will cause cancer in one out of 100 people who drink water containing 50 parts per billion.
- 114 pounds of lead, 4 pounds of cadmium, other toxic heavy metals, and trace amounts of uranium.[49, 50]

When you think of it that way, flicking the switch doesn't seem quite so innocent anymore.

In the following sections, you will find information about how to reduce energy use in your home, and your environmental impact on the road.

Number Four

Home Energy Use

In Jewish tradition, the home is referred to as "mikdash m'at," or "a small sanctuary." We should behave in our home the way we would in any holy place. Being wasteful in our home might be one of the failures we recite on Yom Kippur as we list in the "Al Het" litany the ways in which we did not do our best. Saving energy should be looked upon as doing our share to preserve the resources we were given as gifts by God.

There are so many ways we can save energy at home. In my home, and in my children's homes, I am known as the compulsive energy saver, who constantly walks around turning off lights in rooms which are empty of people.

Recommended Actions:
- Turn off the lights
- Buy energy efficient appliances
- Unplug electronics when you travel
- Adjust your thermostat

Here are some things for you to try. You'll save money and also protect the environment.

It bothers me to no end when I pass, on so many evenings, empty skyscraper office buildings in big cities with all the lights glowing – probably for the cleaning staff who will spend a few minutes in each room. Yet so many people just leave lights on in the home or office all day, not giving any consideration to the cost or negative effect on the environment.

It's so easy just to flick the switch as you exit a room. Children need to be trained to turn off lights when leaving a room; and parents need to model this behavior.

Turn off and unplug appliances, like computers, televisions and other electronic devices when not in use – especially overnight. Even if your appliances are turned off, they are still drawing energy out of the wall, and are a part of the "phantom load" which accounts for over 27 million tons of US annual CO_2 emissions. It can come to up to 15 percent of your monthly energy bill. It is estimated that Americans spend $4 billion a year on electricity for appliances we are not using.[51]

When you buy a new appliance (gas range, refrigerator, dishwashers, telephones, computers, air conditioner, TV, computer, etc.) make sure that it has the Energy Star label (www.energystar.com). This certification insures that the products you purchase are as energy efficient as modern technology permits.

Don't wait until your appliance wears out to get a new one. By replacing older, energy-inefficient products, you'll actually save money in the end.

In selecting a new computer, keep in mind that laptops use over 50% less power than desktops.[52]

A simple, yet important, way to save energy is to replace your light bulbs. Compact fluorescent light bulbs (CFLs) are four times more efficient than standard incandescent ones. They also last longer, both saving money in the long run, and helping to reduce your energy use. (Worried about mercury in your CFLs? See http://www.energystar.gov/index.cfm?c=cfls.pr_cfls_mercury)

Keep your freezer full. Freezers work more efficiently when frozen food is in them aiding in keeping things cold.

Water-efficient appliances, often referred to as "low-flow fixtures," include faucets, showers and toilets, can save water. The processing and treatment of water takes a tremendous amount of energy, so when you save water, you save energy too.

Doing only full loads of laundry, with cold water, also helps. And – guess what - using your dishwasher instead of hand-washing uses less water!

You can also reduce your energy use in heating and cooling:
- Turn down (or up) your thermostat. Every time we turn it down two degrees, we save up to eight percent of our monthly energy bill.

- In the summer, on cool days open the windows and let in cool air, and when it is hot, close the windows and drapes or blinds on the sunny side of the house to keep the cool air in.
- Use ceiling fans to circulate cool air in summer and keep warm air closer to the floor in winter.
- When away from home for an extended period of time, adjust thermostats to 50 degrees during cold months; to 85 degrees in hot months.

For more help on conserving energy, look at these web sites:
- EnergyStar: www.energystar.gov
- The Alliance to Save Energy: http://ase.org/

Put it into Action

Ways I Can Save Energy at Home:

Actions to take now:
_____
_____
_____

Actions I will take in the next six months:
_____
_____
_____
_____

Put a check next to the actions as you complete them.

Number Five

Get an Energy Audit

Rabbi Yosi taught: Woe for the creatures who see but do not know what they see, who stand but do not know upon what they stand. Talmud, Tractate Hagigah 12b

A home energy audit helps you assess how much energy your home consumes, and identify opportunities to make your home more energy efficient. Beyond simple activities like turning off the lights, there are usually good opportunities to make your home more efficient, which are cost-effective as well. Actions like increasing insulation in your home, sealing air leaks or getting new windows can often reduce your home energy use.

> **Recommended Actions:**
> - Evaluate your home's energy use with an energy audit.
> - Learn more at www.energysavers.gov.

The audit can identify problems that may, when corrected, save you significant amounts of money over time. An expert can help you identify the most effective choices for you.

Experts suggest that doing an energy audit, and implementing the suggested changes, can save you as much as 30 percent of your monthly utility bill. That can add up to hundreds of dollars. It's nice to save money while you're doing good for the world.

During the High Holidays, it is traditional to do a *Heshbon HaNefesh* (self-evaluation). You may consider doing another kind of evaluation at this time as well, of the energy you expend in your home. If the High Holidays are too busy a time, consider during it in Elul, the month before, or Heshvan, after the busy days of Tishrei. (Of course, any time is good to do this, but it may be easy to remember at a time that Judaism designates for personal introspection.)

An energy audit can be done by yourself or with a professional. In some states, energy audits are available for free from your energy supplier; in others, professionals can cost several hundred dollars. (Remember that their recommendations may save you a significant amount of money over time.)

To learn more about home energy audits, take a look at Department of Energy's Consumer's Guide to Energy Efficiency and Renewable Energy at http://www.energysavers.gov/your_home/energy_audits/index.cfm/mytopic=11160 The website also includes important guidance for how to select the right professional energy auditor.

- The Department of Energy also offers a "Home Energy Saver" site which enables you to do your own energy audit via your computer, at http://hes.lbl.gov/consumer/.
- You can find professional energy auditors in your state on the EnergyStar website at http://www.energystar.gov/index.cfm?c=home_improvement.hm_improvement_hpwes_partners.
- Some localities offer free energy audits. Check with your local utility company and your state and county departments of energy to see what is available in your area.

A professional will want to see your past utility bills, and will go through every room in the house to check heating/cooling equipment, major appliances, windows, doors, attic and basement (if there is one). The professional will use sophisticated equipment that you likely will not have access to yourself, such as infrared cameras, a blower door infiltration test, etc. The professional will give you a full report, with an estimate of what the cost will be to upgrade your appliances and make repairs to reduce your energy consumption, as well as how much you might save by taking the recommended actions.

Put it into Action!

Energy Audit Completed (Date):

Recommended Actions:

Actions Taken:

Number Six

Green Energy

It is customary in Jewish tradition to make a "b'rakhah," or a blessing, before eating, or before utilizing one of the gifts we are given by God. We make a blessing before putting on a tallit, or before washing our hands. Perhaps we should have a blessing for the fuel we consume that gives energy (food) to our homes, just as we recite a blessing for the fuel (food) which we eat.

As we explained earlier, the energy that fuels our technological world comes mostly from coal, natural gas and oil. Use of these fuels increases carbon dioxide in our atmosphere. Fossil fuels also emit a range of other pollutants which dirty our air and water, and which are already making many people sick (for example, smog which exacerbates asthma, mercury pollution in water which makes some fish unhealthy to eat).

> **Recommended Action:**
>
> Find out if you can buy green power in your state, by checking this website offered by the Department of Energy Green Power Network:
>
> http://apps3.eere.energy.gov/green-power/buying/buying_power.shtml

Most homes in North America are heated and lit by the U.S. power grid, which is connected to our homes. However, the power grid can also supply energy to our homes with resources other than fossil fuels. These include nonpolluting energy sources such as wind and solar energy. In many situations, these green energy solutions are achieved by a simple click of the mouse.

In the United States, more than 600 power utilities in 36 states offer green energy options to residential clients.[53] In some cases, you need only go to the utility company's web site to choose a clean energy option. Clean energy may cost a bit more than fossil fuels but not much more, and as time goes on these rates will become more and more competitive.

Sometimes my friends or neighbors balk at a slightly increased price for renewable energy, but think nothing of spending thousands of dollars on a pricey vacation, or for new bells and whistles on their car, designer clothing, or an entertainment center. Given all the money we spend on luxuries that we don't need, it's worth considering whether a slightly increased cost really is beyond our means, or if we just need to reevaluate our priorities. Perhaps it is time we start recognizing the hidden costs we are billing our children for with our cheap non-renewable energy.

More solar energy falls on the Earth in one hour than the entire world population uses in one year.[54] God has blessed us with a sun that provides so much energy – how grateful we need to be. "God made the two great lights, the greater light to dominate the day and the lesser light to dominate the night...." (Genesis 1:16).

In New York City, for example, a brief visit to www.ConedSolutions.com allows an apartment dweller to pay about $10.00 more per month and change to 100 percent wind power. Switching to a combination of wind and hydropower would cost only about $4.00 more per month.

Folks who live in sunny climates can turn to solar power systems even at your local Home Depot (www.HomeDepot.com). Home Depot is collaborating with BP Solar to offer solar power to residential customers throughout the U.S. There is a free in-home consultation, and Home Depot does all the paperwork, including permits, tax credits and coordination with the power utility. There is a 25-year warranty that comes with installation.

Americans are less than 5 percent of the world's population, yet represented 23% of the world's petroleum use and 19% of the world's coal use in 2006.[55] Since we are clearly using more than our share, we ought to think carefully about ways that will enable us to reduce our fossil fuel use in heating, cooling and lighting our homes.

By being more conscious of our use of fossil fuels in our homes we can become part of the solution to environmental degradation instead of being part of the problem. And at the same time fulfill Jewish responsibility to protect resources for future generations.

Put it into Action

My Average Monthly Energy Bill:

Local Options for Green Power:

Other Action Steps:

Number Seven

Green Your Transportation

In the period before the Industrial Revolution, Jews lived in tight-knit communities where one could walk almost everywhere. Walking to the grocery store, school, synagogue and community center, was the norm.

> **Recommended Actions:**
> - Use a bike or your two feet for short trips!
> - Use public transportation.
> - Consider a car-share.
> - If you need a car, buy one with good fuel economy and keep it well-maintained.

We can't go back to those days when Jews lived in a shtetl. But even if we can't walk everywhere, there are many things we can do to reduce our use of oil. When Maxine and I walk to synagogue on Shabbat, we feel wonderful, not just because we are following traditional Jewish Law, but also because we are reducing our carbon footprint. Reducing oil use has important environmental benefits and also reduces our dependence on foreign oil. Scientists continue to develop low-carbon alternative fuels. We need to continue to work on finding ways to increase the efficiency of cars, trains, buses and airplanes. But the best way is to reduce the amount of motorized travel in which we engage.

On a recent trip to the Far East I saw with my own eyes the horrible congestion of cities in China and Japan. But we don't have to go that far to see bottleneck traffic jams. Anyone who has driven through Manhattan or Los Angeles knows how long it takes to drive one block in rush hour. It's surely easier to walk in such places, if not too far, than to drive. And a lot healthier too!

The Department of Labor reported that in 2009 Americans spent 17.6% of their annual income on transportation, the highest percentage of our income aside from housing.[56] Because of the longer and longer stretch of our national highways system, workers commuting time takes more and more time, leaving less time to spend with the family, to participate in fulfilling community activities, exercise, read, study, and engage in other spiritual activities.

Physical exercise, such as walking instead of driving, is strongly promoted in Jewish tradition. The rabbis pointed out that the verse in Deuteronomy (4:15), "Ve-nishmartem m'od l'naf-sho-tay-khem," "And you shall be very watchful of yourselves...." refers to taking care of your physical health as well as your spiritual well-being. It is a great mitzvah to care for both your soul and your body.

How many of us hop into the car for a short ride to the supermarket, when we can just as easily take a healthy walk to purchases our groceries? Some prefer to ride their bikes to the store, to a meeting, or just to get some good exercise. Some people can choose to live in a small community close to shops, synagogue, schools, and other necessary places of visitation.

As I write this my wife and I are living in Jerusalem for a period of six months. Outside our apartment window, and especially at the market, we see many people pushing what they call here "granny carts," – small wagons on wheels that can hold several bags of groceries. We could take care of our health and the environment by traveling short distances using bicycles and our own two feet.

There are times when walking and biking are simply not a good option for us. But there is still much we can do to reduce the impact of our car usage.

Buy a car with good gasoline mileage, and consider a hybrid car, which can get up to 50 miles per gallon, as opposed to SUVs and other "gas guzzlers," which at present get from 10 to 20 mpg. Some synagogue parking lots are now posting signs to reserve spots near the door for hybrid cars to create role models of those who drive such fuel-efficient vehicles. The American Jewish Committee, which subsidizes staff members who purchase a hybrid automobile, also launched an initiative to reserve parking places in churches and synagogues for fuel-efficient cars. This initiative, says Allyson Gall, former Director of the NJ area AJC says that it "educates people about the issue of our dependence on foreign oil, which affects national security, Israel, human rights, and the environment. Someone might come to a church or synagogue event and see that not just the handicapped get a good parking spot."

While a hybrid car may cost more in the beginning, it will pay for itself after a few years. Some states give financial incentives for those who purchase hybrids. Maxine and I each have a Prius, and find that when it costs us $15. to fill up the tank with gas, our neighbors with SUVs are paying $30 to $50. to do so.

- For a comprehensive review of all hybrid cars available in the United States and an excellent buyer's guide, visit the Hybrid Center of the Union of Concerned Scientists at http://go.ucsusa.org/hybridcenter/incentives.cfm.
- You can also learn about the most fuel-efficient vehicles available at EPA's Green Vehicle Guide at http://www.epa.gov/greenvehicles/Index.do;jsessionid=afb111756544e668189250d0c00ea63d3f67bff95f524d2ff1dac6ba79594ec4.

If you can't buy a hybrid, consider buying a smaller car, and driving it less. Many families can switch from two cars to one. Others can arrange carpools to work, school or shopping.

See www.carpoolworld.com for more information about driving in groups.

If you're going to be in a particular neighborhood see if you can do all your errands there at once, and offer friends to pick something up if you'll be in a store where they may need something. In addition, call the store in advance to see if they have what you need, so that you won't have to go from store to store looking for something.

No matter what kind of car you drive, by taking better care of it you will save money, and burn less gas. Good maintenance includes regular tune-ups, properly-inflated tires (buying a tire gauge will help you measure tire pressure), emptying the trunk of heavy items that cause the car to burn more gas, all can help reduce gas usage.

In more and more cities, people are signing up for "time-share" car rentals, in which each person reserves a car for a certain number of hours a week. It's like a large number of people buying one car and using it when they need it, instead of each person owning their own car. In the USA, carsharing membership rose by almost 120% between 2007 and 2009, thus achieving two major social benefits by reducing the number of vehicles on our nation's roads and the amount of emissions being produced. Research by Frost & Sullivan,[57] professional consultants whose goal is to accelerate growth and develop best practice levels in growth, innovation and leadership, demonstrated that each shared vehicle replaced 15 personally owned vehicles in 2009 — and carsharing members drove 31% less than when they owned a personal vehicle. These two factors translate into 482,170 fewer tons of CO_2 emissions and less travel congestion in urban areas." To learn more, see www.zipcar.com, or www.carsharing.net.

It also saves gas to drive at a steady speed, and avoid sudden braking or accelerating. Drive under 65 or 60 miles an hour.

Don't idle your car more than 30 seconds. For example, turn off the ignition while waiting for your turn in the "drive-thru" at the bank or pharmacy, or in the carpool lane at your child's school. (This applies unless you have a hybrid car, which automatically shuts off the engine when the car is not moving.)

Learn about and take public transportation. Many cities have excellent public transportation systems that enable commuters to get around easily without battling traffic, and some employers even provide financial incentives to their employees to use them. For more details on public transportation in your community, visit http://www.publictransportation.org/.

For intercity travel, consider a bus or train instead of driving or flying. Try to find vacation sites that are closer rather than farther. When you do fly, use carbon offsets (see *#8: Use Carbon Offsets* for more on this).

Better yet, have a meeting on occasion by telephone instead of in person. This saves time, money and gas. A number of good websites offer free or inexpensive videoconferencing tools, including DimDim.com and GoToMeeting.com.

Begin to think "green" when it comes to transportation and you will help the planet become a healthier and safer place to live.

Put it into Action!

Which car trips could I replace with biking or walking?

Which car trips could I replace with public transportation?

How else could I make my transportation more sustainable?

Number Eight

Use Carbon Offsets

Our Torah teaches us in many ways that we must give as well as receive. The agricultural laws in the Torah, such as "Peah" (leaving the corner of the field for the poor, instead of plowing it) and similar biblical edicts remind us of our responsibility to share our blessings with others, and to give back to society in return for all the gifts we get. In the same way, as we use energy products, such as gas and oil, we must think about how to give something back to the planet in return.

> **Recommended Actions:**
> - Consider offsetting your carbon emissions. To learn more, visit:
> - www.carbonfund.org/site/pages/calculator/.
> - To offset carbon with projects in Israel, visit
> - http://support.jnf.org/goneutral/carbonCalc.html
> - http://www.goodenergy.org.il/language/en-US/En/Offset-with-us!.aspx

These days, it's not easy to live a lifestyle based only on renewable energy. To get to work, to travel to visit family, to purchase the food we buy in the supermarket (trucked thousands of miles to us), even to heat our homes, most of us will need to use some non-renewable energy resources. We can, however, "offset" the impact of the carbon dioxide pollution emitted when we take these actions.

By purchasing a carbon offset, we are paying to reduce some of the impact of carbon emissions that we cannot easily avoid. These offsets are used for projects that support renewable energy, save or plant trees, or otherwise reduce carbon emissions elsewhere.

For example, when a person flies from New York to Los Angeles, the plane emits a certain amount of pollution along the way. To offset the pollution, he or she can purchase carbon offsets, according to the number of miles travelled.

Several years ago a young idealistic and altruistic couple, whom I had taught in a summer workshop retreat, invited me to fly to Wisconsin from my home in Princeton, NJ, to perform their wedding. On the wedding program, they explained all the things they were doing to make their wedding celebration environmentally friendly – such as using edible dishes and cutlery (what a surprise that was!). I was extremely impressed, and significantly educated in environmental matters after celebrating the wedding with these wonderful young people.

Among the items on the program which they mentioned was that they calculated the miles all their guests had to travel to attend the wedding (people came from the east and west coasts), and by using one of the standard calculating techniques, they made a contribution to purchase carbon offsets to equalize this expenditure of energy.

There are web sites and calculation devices to help you determine how many carbon offsets one should purchase to "offset" a specific act, such as driving, flying, maintaining a home, etc. One good website is www.carbonfund.org/site/pages/calculator/.

Jewish National Fund has dedicated a significant area within Israel to plant additional trees to offset your carbon usage. You may calculate how many trees will offset your carbon at http://support.jnf.org/goneutral/carbonCalc.html or just go to www.jnf.org/trees.

Also in Israel, the Good Energy Initiative[58] is a unique project established by the Heschel Center for Environmental Learning & Leadership. The GEI works to reduce Israel's greenhouse gas emissions, and to support Israeli energy independence by means of energy efficiency and alternative technologies. The GEI invests its revenues in non-profit social/environmental activities, and it is the only active voluntary carbon offsetting body in Israel. Projects include solar water heaters for low-economy public housing, solar air conditioning in schools in Eilat, and circulating solar-powered medical equipment to children in need. Many Jewish organizations are now offsetting their carbon emissions with the Good Energy Initiative.

To learn more, visit: http://www.goodenergy.org.il/language/en-US/En/Offset-with-us!.aspx

Put it into Action!

Calculating My Carbon Offsets			
Plane Trips this Year	Amount of Mileage	Recommended Offset	Offset? (Y/N)

Section C: Bal Tashkhit: Reduce Waste

"The designer and maker of the earth created the earth not to be a wasteland, but to be lived in"- Isaiah 45:18

"If you believe that you can destroy, then believe that you can fix."

Rebbe Nahman of Breslov

Destroying any physical or material object of use is a violation of Torah law. The law (called in Hebrew "Bal Tashkhit") is based on a biblical verse that prohibits the destruction of fruit trees as a tactic of war. But subsequent Jewish legal tradition, in Talmud, Codes of Jewish Law and other places, extends this prohibition to any act of destruction in peacetime as well as in war.

Here is the Torah source:

When you besiege a city a long time, in making war against it to take it, you shall not destroy its trees by wielding an ax against them; for you may eat of them but you shall not cut them down; for is the tree of the field a human that it should be besieged by you? Only the trees of which you know that they are not trees for food, them you may destroy and cut down that you may build bulwarks against the city that makes war with you until it fall. (Deuteronomy 20:19,20)

Twelfth century scholar Maimonides summarizes centuries of discussion about this law by showing that the law is expanding to cover much more than trees and the use of an ax. He writes: "And not only trees, but whoever breaks vessels, tears clothing, wrecks that which is built up, stops fountains, or wastes food in a destructive manner, transgresses the commandment of *Bal Tashkhit*...."

Rabbi Norman Lamm, Chancellor of Yeshiva University, comments: "What we may derive from this is that the prohibition is not essentially a financial law dealing with property (*mammon*), but religious or ritual law (*issur*) which happens to deal with the avoidance of vandalism against objects of economic worth. As such, *Bal Tashkhit* is based on a religio-moral principle that is far broader than a prudential commercial rule per se, and its wider applications may well be said to include ecological considerations."

Maimonides teaches us that a person should be trained not to be destructive. He gives the example of when a person is buried; the deceased's garments should be given to the poor and not buried in the grave where they are cast to worms and moths. According to Maimonides, anyone who buries the dead in an expensive garment violates the mitzvah of "lo tashkhit."[59] This prohibition moves us towards the more general ethical principle underlying bal tashkhit —that it trains a person not to be destructive.

The Sefer Ha-Hinukh, a thirteenth century text which explicates in detail the 613 mitzvot, elaborates greatly upon the notion of ethical training first mentioned by Maimonides and even sheds light onto why as Jews we should take the mitzvah of bal tashkhit seriously. "The purpose of this mitzvah is to teach us to love that which is good and worthwhile and to cling to it, so that good becomes a part of us and we will avoid all that is evil and destructive. This is the way of the righteous and those who improve society, who love peace and rejoice in the good in people and bring them close to Torah: that nothing, not even a grain of mustard, should be lost to the world, that they should regret any loss or destruction that they see, and if possible they will prevent any destruction that they can. Not so are the wicked, who are like demons, who rejoice in destruction of the world, and they are destroying themselves."[60] Our first instinct should not be to destroy, but rather how we can make use of something with the least amount of damage.[61]

Here is an example of a Talmudic story that is a lovely description of impeccability in not wasting:

Abba Hilkiyah, the grandson of Honi the Circle-maker, would go out to gather wood in the evening. The entire way home, he did not wear shoes, except when he came to a stream of water, and then he would put his shoes on to cross the stream. When he came to an area of thorns and thistles, he lifted the hem of his clothing, exposing his leg to the thorns. When asked about these things, he explained that most of time he could see where he was walking and therefore did not require shoes, but in the water he could not see where he was stepping, so he used his shoes only there to protect his feet. As to why he lifted the hem of his clothing when he came to thorns and thistles, he explained that his leg would heal but his garment would not heal (Tractate Ta'anit 23a). While we may not want to go that far, it is interesting to see how conscious our ancestors were of resource conservation.

What kind of wasteful destruction could be prevented today if we were more careful to observe the mitzvah of Bal Tashkhit? These ideas will be explored in much deeper detail in the next few sections.

Number Nine

Avoid Disposables: Cups and Plates and Much More

It's so much easier to use disposables – no cleaning, washing, drying, storing, etc. Yet they are a major cause of pollution in modern society.

Are you a coffee drinker? In 1999 there were 108 million coffee consumers in the United States.[62]

> **Recommended Actions:**
> - Use Cloth Bags for your purchases.
> - Carry around a reusable bottle, cup, or utensils so that you can reduce plastic waste.
> - Use cloth napkins and rags instead of paper towels and paper napkins.

Many of those consumers use disposable cups. But we can reduce the waste from disposables. For example, Starbucks, in partnership with the Environmental Defense Fund, has developed an approach to reduce waste by using 10% recycled paper for their cups, and by using a "sleeve" to protect fingers from hot temperatures, instead of a double cup. Starbucks also offers a 10-cent discount to customers who bring in their own mug.[63] You can take this on and also encourage your local coffee store to do the same.

Those who drink water or coffee can carry around their own ceramic mug or stainless steel travel cup. Some people even carry around their own plates and utensils. See http://www.to-goware.com/ for a sustainable solution. These make great gifts for your eco-conscious friends!

We use between 500 billion and a trillion plastic bags[64] each year in the United States, in supermarkets, pharmacies, book stores, department stores, clothing stores, and others. Many of the disposable plastic bags make their way from trash bins into the oceans or rivers and kill fish, birds, and other wildlife[65], where they will take 10-20 years to decompose.[66] Many others end up in landfills, where they will be around for a lot longer.[67]

When I check out at the supermarket, I am typically asked, "Paper or plastic?" My standard answer is "No." I then show the checkout person my own cloth bag(s) that I brought specifically so that I would not be guilty of using more disposable bags of paper or plastic. If I have time, I give the person a short "lecture" on the harm of using plastic bags (which mostly they appreciate). My wife, Maxine, who is known as a passionate Israel activist, adds, when she shops, that she doesn't want to enrich Saudi Arabia – since plastic comes from oil. Parts of Los Angeles County have banned plastic bags in stores, and one would hope that other communities and states will follow.

Cloth grocery bags are available everywhere. They are very inexpensive, reusable, and avoid the problem of disposables. Some stores will give a small rebate to those who bring their own cloth bags. Still other stores, especially grocery stores, collect plastic bags for recycling—even ones not from their store.

Sometimes people tell me, I'd like to bring a cloth bag but I always forget it at home! The best solution Maxine and I have found is to keep the bags in our car. The cloth bags come into the house, we empty them, and they go right back out to the car. That way, we have them when we go shopping. There are also small bags that fold up into your purse or backpack and can easily be unrolled and used at a moment's notice. For an example, see http://www.chicobag.com/.

Some other simple ideas for reducing waste in your house:
- Use real dishes and cups. It may feel more convenient to use plastic or paper cups and dishes, but the long-term impacts are anything but convenient.
- Use cloth napkins in your home. They don't take much space in a full laundry load, and you can avoid mounds of paper waste. They are also more festive for nice meals.
- Instead of paper towels and "cleaning wipes," use rags (otherwise known as old clothing or towels that you are not using anymore) for cleaning.
- Instead of paper towels or napkins, use hand towels in your bathroom.

Of course, these days most electronics are powered by rechargeable batteries that are plugged into the wall. If you still use batteries (for example, for children's toys), choose rechargeable ones if possible. If your electronic device requires more energy than a rechargeable battery can reliably deliver, you can recycle disposable alkaline batteries at local stores or municipal collection sites. (See Earth911.com for recycling sites.)

As one author[68] put it starkly: Disposable is (literally) a dirty word!

> **Put it into Action!**
>
> List all the disposable products you use here:
>
> _____
> _____
> _____
> _____
> _____
> _____
> _____
> _____
> _____
>
> *Place a star (*) next to those you will reduce in the coming month.*

SECTION C: BAL TASHKHIT: REDUCE WASTE   49

Number Ten

Reducing the Impact of Electronics Waste

In today's world of "planned obsolescence," all of us have a tendency to toss broken or out-of-date electronic gadgets and machines, such as cell phones, computers, TVs, iPhones, etc. in the trash. We sense it is better and easier to buy a new one rather than try to repair the old one.

In the commercial world, many industries intentionally produce gadgets and machines that will become outdated quickly so they can sell hungry consumers newer, faster, more powerful, prettier and shinier ones. This is known as "planned obsolescence."

> **Recommended Actions:**
>
> When you are finished with your electronics and inkjet cartridges, don't throw them in the trash! Ask:
>
> - Is there a non-profit organization that is collecting them?
> - Is there a local collection coming up in your area?
> - Can you recycle them with a local store, a website, or your post office?

Americans own nearly 3 billion electronic products.[69] For each new music device, e-reader, computer, or smart phone that is developed, one or more becomes outdated or obsolete. As a result, we're storing or discarding older electronic products at an increasing pace.

In 1998, studies estimate about 20 million computers became obsolete in one year. In 2005, EPA estimated that between 26-37 million computers became obsolete. Along with computers, TVs, VCRs, cell phones, and monitors—an estimated 304 million electronics—were removed from US households in 2005, with about two-thirds of those still in working order, according to Consumer Electronics Association (CEA) estimates.[70]

According to the EPA, 2.25 million tons of TVs, cell phones and computer products were ready for end-of-life management in 2007. Yet, only 18%

(414,000 tons) was collected for recycling. The rest, 82% (1.84 million tons), was disposed of, primarily in landfills.[71]

Whether they are still functional, or just carrying useful metals which can be recycled and reused, tossing these older gadgets and machines in the trash violates the spirit of "Bal Tashkhit." It also does serious damage to the environment. Typical computers contain lead, aluminum, gallium, vanadium, beryllium, chromium, cadmium, mercury, arsenic – chemicals that are known to cause serious health effects.[72] According to the National Institutes of Health, the glass cathode ray tubes (CRTs) which are found in televisions and computer display monitors each contain an average of four pounds of lead.[73]

Dumping cell phones, computers, TVs and other modern technological devices in landfills is called by the technical term "e-waste," and is outlawed in 20 states in the United States. Yet preventing e-waste in our landfills can create an export stream leading to environmental damage in other countries.[74]

So, what should you do with your used electronics? Reusing or donating them is the best option. Suggestions for reuse and donation options can be found from the Environmental Protection Agency at http://www.epa.gov/epawaste/conserve/materials/ecycling/basic.htm#reuse.

If donation for reuse or repair is not a viable option, recycle the products rather than putting them in the trash. You can find more information and suggestions for recycling electronics at http://www.epa.gov/epawaste/conserve/materials/ecycling/basic.htm#recycling.

Another way to reduce your impact is to choose new equipment that is environmentally preferable. You can encourage electronics manufacturers to design greener electronics by considering green criteria as part of your purchasing choices.

Households, companies, and governmental organizations can encourage electronics manufacturers to design greener electronics by purchasing computers and other electronics with environmentally preferable attributes and by requesting takeback options at the time of purchase. Look for electronics that contain fewer toxics, use recycled materials, are energy efficient, use less packaging, are designed for easy upgrade or disassembly, and offer leasing or takeback options.

You can also look for products that meet performance criteria to show that they are environmentally preferable. For electronics, the Electronic Product Environmental Assessment Tool (EPEAT) provides an opportunity for manufacturers to secure market recognition for efforts to reduce the environmental impact of its products. EPEAT is a program of the Green Electronics Council, a volunteer advisory board whose membership is comprised of environmental advocates, institutional purchasers, manufacturers, government policy professionals, researchers and electronics recyclers.

EPEAT currently covers desktop and laptop computers, thin clients, workstations and computer monitors. Desktops, laptops and monitors that meet 23 required environmental performance criteria may be registered in EPEAT by their manufacturers in 40 countries worldwide. Registered products are rated Gold, Silver or Bronze depending on the percentage of 28 optional criteria they meet above the baseline criteria. EPEAT operates an ongoing verification program to assure the credibility of the registry. To search by manufacturer or by criteria, visit http://www.epeat.net/.

Resources to help you:

To find out how to dispose of electronics in your area, visit http://earth911.com/ and search for recycling centers for "Electronics" in your area.

Canfei Nesharim, the Jewish environmental awareness organization (www.canfeinesharim.org), also provides a list of organizations that take electronics, compact fluorescent light bulbs, cellular phones, inject and ink cartridges at http://canfeinesharim.org/learning/holidays.php?page=19961.

Many local non-profits collect cell phones, ink cartridges and other electronics. The organization receives a small donation for each item recycled, and the products are either re-used or disposed of properly. Ask around in your neighborhood to see if you can find one that would be happy to recycle your products for you.

Some local municipalities organize "electronic waste collection days" several times a year, so check with your locality to see if one is coming up.

The U.S. Postal Service in 2008 also launched a free service that allows customers to recycle small electronics and inkjet cartridges by mailing them free of charge. Customers use free envelopes found in 1,500 Post Offices to mail back inkjet cartridges, PDAs, Blackberries, digital cameras, iPods and MP3 players – without having to pay for postage. According to the Postal Service press release, postage is paid for by Clover Technologies Group, a nationally recognized company that recycles, remanufactures and remarkets inkjet cartridges, laser cartridges and small electronics. If the electronic items or cartridges cannot be refurbished and resold, their component parts are reused to refurbish other items, or the parts are broken down further and the materials are recycled. Clover Technologies Group has a "zero waste to landfill" policy: it does everything it can to avoid contributing any materials to the nation's landfills.[75] Check to see if your Post Office participates; if so, the envelopes will be available to you there.

At Wirefly.com, you can trade unwanted working electronics for cash delivered to a PayPal account or a charity. Learn more at http://www.wireflytradeins.com. A similar program—with free shipping—offers gift cards via TigerDirect.com. Learn more at http://tigerdirect.cexchange.com/.

Some retailers also have electronics trade-in programs that provide store gift cards. They include Costco, Best Buy and Sears. Most operate the same way. You select the trade-in item from a menu, fill out the fields on its condition and accessories, and get an instant trade-in value.

Much useful information can also be found at additional websites, including:
- Electronics Take Back: www.electronicstakeback.com
- the EPA web site: http://www.epa.gov/epawaste/conserve/materials/ecycling/index.htm
- E-stewards http://e-stewards.org/, an electronics recycling program recently recognized by the EPA.

Put it into Action!

Electronics I need to recycle, reuse or donate this year:

Items to reuse or donate:

Items to recycle and where I will recycle them:

Electronics I plan to purchase this year:

Ideas for buying more green electronics:

Number Eleven

Drink Tap Water Instead of Bottled Water

I will be standing there before you on the rock at Horev. Strike the rock, and water will come out of it, that the people may drink. And Moses did so in the sight of the elders of Israel.
Exodus 17:6

Behold, God is my salvation! I will trust, and not be afraid: for Ado-nai is my strength and *my* song; he also is become my salvation. Therefore with joy shall you draw water out of the fountains of salvation.
Isaiah 12:2-3

Ho, all who are thirsty, come to the waters....
Isaiah 55:1

Scientists report that in recent years Americans spend $11 billion per year on bottled water. That translates into nearly 9 billion gallons of bottled water per year in the US alone. Worldwide usage is over 52 billion gallons per year.[76]

Recommended actions:
- Learn about the quality of your drinking water.
- Replace your bottled-water use with tap water, filtered if necessary.
- If you must use a disposable plastic water bottle, don't store it in a hot place or for a long time. Drink, and recycle when done.

In several ways this is very bad for the environment.

First of all, most of this water has to be transported from city to city, employing ships, trains and trucks across the country, using gasoline and increasing carbon output in the atmosphere. We are told that a billion bottles of water per week are transported![77]

Second, the plastic bottles are very harmful to the environment. Plastic is made from oil, a non-renewable energy source (as we've discussed earlier).

SECTION C: BAL TASHKHIT: REDUCE WASTE

The amount of oil it takes to create the plastic bottles that are used to sell water could run 100,000 automobiles for a whole year![78] We throw away 80% of these bottles, rather than recycle them. These plastic bottles fill our landfills and often end up in our waterways.[79]

Third, by drinking water from plastic bottles we are wasting an enormous amount of money. People have gotten the crazy idea that bottled water is better than tap water. Yet the opposite is true! The fact is that low-cost water from your tap is often more healthy than expensive bottled water. According to experts, much of the bottled water we purchase in supermarkets and elsewhere is simply purified tap water, and costs over 1000 times as much as what comes out of the faucet.[80] For a person to buy enough bottled water to drink the recommended 8 to 10 glasses a day, it costs about $3000 a year. For a similar amount of tap water, the cost is about one dollar! Furthermore, whereas all municipal water systems are required by law to publish water quality tests annually, only 18% of bottled waters disclose quality reports that include contaminant testing results.[81] Both Aquafina and Dasani have acknowledged that their product is nothing more than bottled tap water.[82]

How does a responsible Jew avoid wasting money, polluting the environment, and using plastic (made from petroleum, which is not a renewable source of energy)?

- Learn about your drinking water. Request a copy of "Consumer Confidence Report" from your Water Utility Co. to find out where your water comes from and contaminants found: www.epa.gov/safewater/dwinfo/. Some drinking water is just fine to drink as it is.
- If you have concerns about your drinking water, filter your water at your tap. Compare filters for contaminants removed, costs/gal, filter replacement frequency, and efficiency at www.waterfiltercomparisons.net. The "Environmental Working Group" has a web site with a water filter buying guide: http://www.ewg.org/tap-water/getawaterfilter.
- Use safe, re-usable bottles like stainless steel or glass.
- If you must use a disposable plastic water bottle (for example, if you have run out of water on a long trip), don't leave it in your car, and don't store it in hot places, or for long periods since the plastic leaches into the water. When done, recycle it!

- Buy colored "milky" plastic rather than "clear"; the hard translucent plastic (identifiable by the recycling symbol #7) is often made with BPA. Water bottles with recycling symbol #1 are meant to be used once only.
- Don't wash & reuse plastic bottles. Microbes grow easily; they are hard to clean, easy to scratch and chemicals may leach.
- Talk to your office manager about companies that can provide safe, cost-effective water filtering at your workplace (one example is www.uscoffee.com/content/water.asp).[83]

Put it into Ation!

Drinking Water Quality Information: (Date:_____)

Water Filter Options to Consider:

Actions Taken:

Section D: Reduce, Reuse, Recycle

Live simply, so that others may simply live.   Gandhi.

"The earth belongs to Ado-nai, and all it holds, the world and its inhabitants." Psalms 24:1

In an ancient book in the Mishnah, Pirke Avot (Ethics of the Fathers), there is a popular aphorism by Rabbi Shimon ben Gamliel: "On three pillars the world rests – on the Torah, on Divine Worship, and on Acts of Kindness" (1:2).

To borrow that metaphor, environmental responsibility rests on three pillars: Reduce, Reuse, Recycle.

There is more than an accidental metaphoric comparison between these two dicta. As we improve our environment, we are following the dictates of the Torah (to honor God's beautiful world), Divine Worship (thanking God for the gift of our universe by honoring it and protecting it), and Acts of Kindness (handing to our children a sustainable world). "Good people leave an inheritance to their grandchildren" (Proverbs 13:22).

Let's examine each of these three "pillars," and see how we can improve and honor our world by employing them.

Number Twelve

Reduce

"Take care...lest you eat and be satisfied and you build good houses and settle, and your cattle and flocks increase, and you increase silver and gold for yourselves, and everything that you have will increase--and your heart will become haughty and you will forget the Lord your God..." Deuteronomy 8:11-14

Rabbi Adin Steinsaltz, whom many consider the greatest Jewish scholar of the modern age, wrote in an essay called "The More Things Change, the More They Stay the Same,"

The impoverished citizens of today's world may be far less poor than they were 1,000 years ago, but their envy of those who have more has not changed much.
Today's rich – as rich people always have – find that a more comfortable life is not necessarily a happier one.
An expensive gourmet dish in a lavish restaurant will never be as tasty as a meal eaten after two days of fasting.
The joys of a very posh wedding will never be as satisfying as the smile of someone we love.

The ancient book of Jewish ethics, *Pirke Avot* (*Ethics of the Fathers*) [2:8], gives this wise advice: "The more possessions, the more anxiety." In modern times, we can easily relate to this prudent guidance. The more we have, the more time and money we must devote to buying, fixing, calling repair experts, updating, and so on.

There are many unhealthy results from putting too much emphasis on *things*. In Jewish tradition, idolatry is often

Recommended Actions:
- Before you buy something new, consider whether you really need it.
- Reduce your junk mail and cancel unwanted magazines, papers, catalogs, etc.
- Rent or borrow items you do not use often.

defined by placing high priority on unimportant things, and lower priority on important things. When material possessions become the focus of our lives they can indeed turn into a form of idolatry.

So, consider simplifying. Before you go out to buy the new gadget, think: do I need this? Will I use it? Will it really improve my life? Could I wait 1-2 years for the next model? This type of thinking can reduce all of our resource use over time. Our family usually keeps our automobiles until they run into the ground. We ignore the appealing advertisements that bombard us daily to buy or rent the newest model with the newest bells and whistles.

There are so many ways to *reduce* the number of our possessions, and the material goods we utilize in our daily routine. My synagogue, The Jewish Center of Princeton, NJ, has recently offered members the option of receiving the monthly synagogue bulletin by email attachment instead of a hard copy in the mail. This is the kind of step synagogue members can easily push for in their own community.

I also have arranged with my bank and credit card companies to "go paperless." Almost all my financial dealings and records are now found on my laptop. Going paperless saves an enormous amount of paper, saves trees, and unclutters our offices and our lives.

Instead of buying scratch pads, I neatly cut pieces of paper that I have already used, and turn them over to the other side for scratch paper. I even use the back of envelopes that bring me mail, for scratch paper.

For decades, I was an avid collector of books. I know, therefore, how precious it is to have one's own personal library. To this day, I still buy important books, especially reference books. As a rabbi, I can find no fulfilling substitute for owning copies of the Bible, Talmud, and other "Sifray Kodesh," holy volumes containing the sacred words of our tradition.

However, I have also found that book buying can become an addiction, just like buying too many articles of clothing. When I retired from being an active synagogue rabbi, I gave away hundreds of books to young rabbinical students – which gave me much joy. About half of the books I had never had time to open, and I would have saved lots of paper, money and space by buying fewer books.

Consider using your local library instead of buying every book you want to read. Most libraries today also have movies, booktapes, DVDs and other items that you can borrow and return for free instead of buying new ones and cluttering up your home shelves. Or join one of the many companies that provide films by mail or online, instead of buying movies on DVD that you'll view once and then will sit on a shelf forever. Especially today, one can also read many audio books on CDs, and the number of books available on an e-reader such as a Kindle, or other non-paper devices is increasing daily.

As an observant and committed Jew, I can never bring myself to say that a Jewish home should not possess a library of important Jewish books. On the other hand, one should also remember that much reading and research can be done on the internet and other computer software programs.

Since I am a writer, and have published almost forty books, I now try to print manuscripts on both sides of a page, to save paper. This is just one small example of how we can be creative and save paper.

I have also cancelled most of my magazine subscriptions, newsletters, and newspaper, and now read the same things on-line, or while waiting in the dentist's or doctor's office – or in the barber shop. Those who prefer to read a hard copy can at least recycle them, but reducing is preferable. This site makes it easy for you to avoid receiving catalogs and junk mail: www.catalogchoice.org.

The web site www.reduce.org suggests purchasing products that are returnable, reusable or refillable, and gives an abundance of other examples of how to *reduce* the amount of stuff we fill our homes and live with. Another example the site gives is to rent or borrow items that you do not use often, such as chairs, tables, ladders, power tools, etc.

For an article about a report by Cornell University on how objects do not make us happiest, see Figure 5.

For extra motivation in this realm, see the informative video, "The Story of Stuff," by Annie Leonard (www.storyofstuff.com). The web site also contains interviews with Annie Leonard by leading TV personalities.

Put it Into Action!

Purchases that will not really add to my happiness:

Junk Mail I can reduce:

Actions Taken:

Figure 5

March 31, 2010
Study: Glee from buying objects wanes, while joy of buying experiences keeps growing

By George Lowery

The satisfaction we get from buying vacations, bikes for exercise and other experiences starts high and keeps growing. The initial high we feel from acquiring a flashy car or megascreen TV, on the other hand, trails off rather quickly, reports a new Cornell study.

Why are experiences more satisfying? For one thing, it's harder to compare them to others' experiences; they belong to us alone.

"Your experiences are inherently less comparative, they're less subject to and less undermined by invidious social comparisons," said professor of psychology Thomas Gilovich, who published the study with Travis J. Carter, Ph.D. '10, in a recent issue of Journal of Personality and Social Psychology.

People are less satisfied with material purchases because they are more likely to second-guess what they could have had (such as a new model or a better price), the researchers found. Consumers spend more time thinking about material purchases they didn't choose than they spend when they buy an experience.

"There's a lot of work in the area of well-being and happiness showing that we adapt to most things," Gilovich said. "Therefore, things like a new material purchase make us happy initially, but very quickly we adapt to it, and it doesn't bring us all that much joy. You could argue that adaptation is sort of an enemy of happiness. Other kinds of expenditures, such as experiential purchases, don't seem as subject to adaptation."

Gilovich conducted studies about five years ago that found people get more enduring happiness from their experiences than their possessions. The new research looks at why that is.

"Imagine you buy a flat panel TV. You come to my house, and I have a bigger, clearer picture than yours. You're bummed out," Gilovich said. "But suppose you go on a vacation to the Caribbean. You find out I've done the same, and mine sounds better than yours. It might bother you a little bit, but not nearly to the same degree because you have your memories; it's your idiosyncratic connection to the Caribbean that makes it your vacation. That makes it less comparable to mine, hence your enjoyment isn't undermined as much."

In one study, a bag of potato chips and a chocolate bar were both on a table. The volunteers were told they could have the chips, while the researchers implied that others got the chocolate. Another group of participants received a small physical gift that was next to a better gift that was intended for someone else. The participants reported they felt less satisfied in the latter case.

"Visible comparison undermined the enjoyment of the material goods, but it didn't undermine the enjoyment of the experiential good [potato chips]," Gilovich explained. "If you consume an experience in the presence of something better, it doesn't as consistently or powerfully undermine the experience."

What does it all mean? "Our results suggest that if people get more enduring happiness from their experiences than their possessions, at a policy level, we might want to make available the resources that enable people to have experiences. You can't go hiking if there are no trails. And if those are the kinds of things that give people more enduring enjoyment, we need to make sure we're creating the kinds of communities that have parks, trails and so on that promote experiences that produce real enjoyment."

The study was funded by the National Science Foundation.

Figure 6

Simplicity as a Jewish Path
Reconstructionism Today Volume 10. Number 2.
RECONSTRUCTIONISM TODAY

a voice for creative Jewish living

VOLUME 10 NUMBER 2
WINTER 2002/2003

Simplicity as a Jewish Path
By Moti Rieber and Betsy Teutsch

Contemporary American life is characterized by relentless stress and rampant consumerism. For many of us, the pursuit of stuff — its purchase, storage, maintenance and disposal — actually works against quality-of-life as measured by health, happiness, and feelings of fulfillment.

In response, the burgeoning Voluntary Simplicity movement advocates cutting back on personal consumption in order to achieve a better balance of time, money, and material possessions. Voluntary Simplicity adherents present a simple formula: Consume less to have more time, more money, and an environmental dividend. One can work fewer hours and expend less time on shopping and maintenance of possessions, while saving money and resources.

Old-fashioned frugality may seem a Depression-era relic, but Voluntary Simplicity is not about poverty or deprivation. Studies show that once our basic needs for food, shelter, and attachment are met, happiness is dependent on a loving family and friends, good health, meaningful work (paid or unpaid; what matters is that it gives us a sense of worth), involvement in a larger community, and spiritual expression (see *The Pursuit of Happiness* by David G. Myers). In other words, higher income and what it buys turn out to be pretty much absent as contributing factors in human happiness. Having a house twice as large does not double your satisfaction in life. Simplicity is about discovering what is "enough" in your life — based upon thoughtful analysis of your lifestyle and values — and letting go of the rest.

Jews have a long and complicated relationship with money. On the one hand, the Jewish tradition is replete with anti-consumerist messages. The Torah tempers the tendency to worry too much about financial security with such legal imperatives as *tzedakah* (charitable giving), Shabbat (taking one day a week completely off), and the sabbatical year (allowing the land to lie fallow every seventh year). Rabbinic literature expresses notable reservations about materialism, as in *Pirke Avot*, where Hillel teaches that "the more possessions, the more worry" and Ben Zoma teaches, "Who is rich? The one content with his or her portion." This wisdom perceives that human happiness flows from state of mind, not from an abundance of material possessions. In the medieval period, some Jewish communities enacted sumptuary laws limiting excess in dress, food, and festivities, in order to decrease competitive ostentation. The simplicity of Jewish burial rites reflects similar concerns. In modern times, the kibbutz movement was founded on an ideal of simple living and anti-materialism.

On the other hand, most American Jews are descendants of immigrants who worked hard at menial jobs in an effort to get their foot in the door, attain the security that America offered, and provide an education for their children. The progression from sweatshop to City College to suburbs is familiar. Often, conflicting attitudes towards money made the trip as well. Rabbi Mordechai Liebling, Torah of Money Director for the Shefa Fund (and former executive director of the JRF), often explores such conflicts in workshops for funders. "While Jews have become the wealthiest ethnic group in America per capita," he says, "the centuries of Jewish poverty and oppression have left a residue of insecurity, anxiety and even shame about financial security."

In a sense, Jews are "hard-wired" to expect fear and scarcity. We rarely question the trade-offs made for the sake of upward mobility. Even if we have inklings of something being out of whack ("When there's too much of something," says one Yiddish proverb, "something is missing"), we may lack the support, insight, nerve — and even time! — to address the issue.

***Still, Jews interested in improving their quality of* life** by cutting consumption need not look far afield. There is a philosophy as well as a practice of Voluntary Simplicity (the Hebrew term is *histapkut b'me'ut* — contentment with less) embedded in Judaism's unique effort to balance private, individual behavior and communal relations. Applying the Jewish values model developed by Dr. David Teutsch [*A Guide to Jewish Practice: Attitudes, Values and Beliefs,* available from the RRC] to issues of time, money, and consumption, we discern seven essential principles of Jewish Voluntary Simplicity:

1. ***Anava* (humility).**
   We are instructed to walk humbly with God. This suggests that we might need to contract ourselves and take up less space. Environmentalists talk about minimizing "ecological footprints." One way to do this is by choosing to live beneath our means. Examples include eating a vegetarian diet, buying fewer clothes, and living in more modest quarters. Driving a hybrid (electric-gas) car instead of an SUV reduces the ecological footprint by two thirds. *Anava* means that we are not entitled to a hugely disproportionate share of the planet's resources, even if we have the wealth to pay for it.

2. ***Ho'da'ah* (gratitude).**
   This value is largely absent from our commercial culture. Realizing that what we have is a gift, not an entitlement, is a spiritual discipline. Training ourselves to be satisfied with what we have, and shutting down our wish lists for more, can be culturally subversive.

Throw out your catalogues! Take a moment to say a blessing before you eat. As contemporary simplicity philosopher Jerome Segal puts it, "Consider the act of saying grace before a meal. Here the core is an attitude of thanksgiving, of appreciation. The focus is on recognizing the full value of what one has, rather than lamenting what one does not. While one can mouth the words, one cannot authentically begin a meal with a benediction of grace and at the same time maintain a sense of dissatisfaction with what one has. There is a certain peaceful contentment that is part of genuine thankfulness."

3. *B'al tashkhit* (**avoiding waste**) and
haganat hatevah (**preserving nature**). American life is characterized by excess: Our houses, cars, and even our bodies are getting bigger and bigger. If all the world consumed at our level, it would take four planets to meet the demand.

Find ways to avoid waste in your personal life. Stop wasting food. Use compact fluorescent light-bulbs, which last nine times as long.

4. *Bitul z'man* (**wasting time**).
The moments of our lives are precious, and we don't know how many there will be. Creating an alignment between our values and how we spend our time is an option denied those billions of people, past and present, for whom survival is a struggle. Most Jewish Americans, however, do have choices. Do we want to work long hours to maintain an opulent lifestyle? Do we want to spend six hours a week shopping, as the average American does? How much time are we willing to spend in a car, running errands and commuting? How do we find time to nurture ourselves, let alone support partners, family, friends and community? *Your Money or Your Life*, a Voluntary Simplicity classic by Vicki Robin and Joe Dominguez, asks the question: How much of your time/life force does it take to buy things? Is it worth it?

5. *Tzedek* (**justice**) and *tikkun olam* (**repair of the world**).
Voluntary Simplicity helps us to free up time and money to devote to these *mitzvot*. We are commanded to give *tzedakah*, which is an obligation, not an option. This is a unique aspect of Jewish Voluntary Simplicity, compared to the more privatistic American model. The standard for giving that the Bible sets is ten percent. Our checkbooks would look very different if we met this standard — yet with careful consumption, it might be achievable.

Tzedakah can also be given through divestment of excess stuff that is useful to others. Adopt a practice to give away clothes every time you buy new ones. The blessing "Praised are You, God, who clothes the naked" can be said both when acquiring and when giving away clothing.

6. **Kehillah (commitment to community).**
In *Bowling Alone*, Robert Putnam highlights the effects of the decline of civic engagement in America: alienation, isolation, even depression. We Jews have a long, powerful tradition of living in community, and understand the crucial relationship between individual and community. In modern times, community association is voluntary, and Jewish communities must work hard to remain healthy. Informal systems of connection within the community help.

Try using your synagogue listserv as a sharing tool, matching people who need things with people who have and don't need them. Our *minyan* has exchanged a remarkable flow of things — medical equipment, bikes, inkjet cartridges — just by posting them online. Instead of spending time and money shopping for the items, community members spend time swapping. You might also create a *shul* support group to share ideas and strategies for dealing with consumerism.

7. **Menuhah (rest and renewal).**
Every year, there is an annual effort to highlight overconsumption by turning the Friday after Thanksgiving, traditionally the biggest shopping day of the year, into "International Buy Nothing Day." Jews have inherited a tradition that sponsors one of these days every week!

Shabbat, a day of cessation from commercial transaction, is a cornerstone of Jewish life. But Shabbat is not only about avoiding work or not consuming — it is about getting off the economic treadmill and facing each other as people rather than as economic actors. Shabbat is about deciding for a day to let everything, including ourselves, just be. These fifty-two annual "Buy Nothing Days" allow us to trade consumption for personal and communal renewal.

The seven values of Jewish Voluntary Simplicity are embodied in certain positive contemporary developments in the Jewish community. For example, in response to the tendency toward bigger and gaudier *b'nai mitzvah* and weddings, a "neo-sumptuary" literature has emerged, typified by Jeffrey K. Salkin's *Putting God on the Guest List*, a guide to increasing the spiritual significance of these occasions. This approach encourages giving *tzedakah* commensurate with the celebration's cost (MAZON, the Jewish community's premiere hunger-relief organization, requests three percent).

In the Orthodox world, a group of rabbis have gone further and created "The Guidelines," sumptuary laws for today. Concerned that members of their communities are going into debt to create celebrations that "keep up with the Goldbergs," the guidelines stipulate maximum numbers of guests and musicians and even specify appropriate centerpieces, flowers and menus, regardless of the celebrants' incomes. Rabbis who sign on refuse to perform non- Guidelines weddings. In the liberal community, which lacks enforcement, downsizing will have to come through brave role models and communal processing.

Another traditional tool for reducing duplicative consumption, the *gemakh*, could be recon- structed for our contemporary situation.

A *gemakh* (the word is made from the first letters of *gemilut chasadim*, deeds of lovingkindness) is a communal lending system. Typically, one person takes responsibility for collecting used items in good condition — everything from wedding clothes to computers — to lend them where needed. In many communities, this is already done informally, as when you hand off maternity clothes to a pregnant fellow-congregant, and she in turn does the same.

A fancy-clothing *gemakh* would save harried parents of *b'not mitzvah* many trips to the mall. These party clothes are expensive, yet are only worn a few times before their owners either outgrow them or outgrow the bar/bat mitzvah circuit. By collecting dresses, shoes, accessories, ties, suits, etc., and making them available to the next year's crop of kids, much would be gained. Such a *gemakh* would create a communal culture that de-emphasizes shopping, and the money not spent on a fancy outfit could be donated to *tzedakah*.

If Simplicity is both authentically Jewish and sensible, why do we feel ambivalent about downsizing? Why do we not all flock to live simpler lives?

One reason is that Jewish life in America is very expensive, involving synagogue dues, JCC memberships, charitable donations, bar/bat mitzvah expenses, trips to Israel, ritual objects and, for parents, tuition for supplemental or Jewish day school, Jewish summer camp, and youth movement trips. Of course, we are not advocating that people drop out of Jewish life in the process of simplifying!

Another complicating factor for those who are parenting in this hyper-commercialized atmosphere is the resistance to limiting consumption that comes from kids themselves, and from the expectations of affluent Jewish life: music and sports lessons, entertainment, vacations, private school tuition for many, and a stream of new clothes, electronics, toys and sports equipment.

In *Blessings of a Skinned Knee*, psychologist Wendy Mogel opines that Jewish values should work to limit these expectations, which are in the long run destructive for children, who need limits. In an environment drenched with advertising, it is difficult for parents to resist pressure from their kids, as well as the messages from society telling them they must give their children every competitive advantage. The real competitive advantage, however, we can give our children is healthy values and a vibrant community.

Simplifying one's life is a long-term process, done most effectively with the support of others who are doing the same. Cecile Andrews has popularized the concept of a "simplicity circle," which meets with a facilitator over time to tackle these issues and to share experiences, ideas, frustrations and successes. A Jewish simplicity circle might have a greater emphasis on the spiritual disciplines and rewards inherent in the simplifying process, using

Jewish vocabulary and placing focus on community affairs. Members might also become effective advocates for adopting synagogue policies compatible with simplicity.

Some of our suggestions are individual, accomplished by modifying personal habits; some are communal, and may take some measure of education and group building. They will all help counter the rampant consumerism in American society, and do so within a demonstrably Jewish framework. We are interested in helping to launch such initiatives and have started a website, *JewishSimplicity.org*.

Images by Lawrence Bush

For Further Reading

- A new web site edited by the authors: *jewishsimplicity.org*.
- *The Circle of Simplicity: Return to the Good Life*, by Cecile Andrews
- *Bobos in Paradise: The New Upper Class and How They Got There*, by David Brooks
- *Jews, Money, and Social Responsibility*, by Lawrence Bush and Jeffrey Dekro
- *Affluenza: the All-Consuming Epidemic*, by John de Graaf
- *The Blessing of a Skinned Knee*, by Wendy Mogel
- *Bowling Alone*, by Robert D. Putnam
- *Graceful Simplicity: Toward A Philosophy and Politics Of Simple Living*, by Jerome M Segal.
- *Tightwad Gazette*, by Amy Dacyczyn
- *The Overspent American*, by Juliette Schor
- *The Simple Living Guide*, by Janet Luhrs
- *www.thegarden.net/simplicity* is a supersite of Simplicity sites. Also check out: *newdream.org*.

From the Winter 2002/2003 issue of the JRF Quarterly ***Reconstructionism Today.*** © 2002/2003 by Jewish Reconstructionist Federation (JRF). All rights reserved.

Rabbi Moti Rieber (RRC 2004) serves Congregation Beth Shalom in Naperville, IL.

Betsy Platkin Teutsch is the artist/calligrapher of the Kol Haneshamah prayerbook series.

Number Thirteen

Reuse

In my family, we have always had the custom of handing down clothing from older siblings or cousins or friends, to younger ones. We have pictures on our walls of cousins and friends wearing the same clothing that our children wore when they were young. This is a good example of *reusing*.

There are at least two ways to *reuse*. First, you can take something that was created for a single purpose– like an empty bottle – and put it to use in a new way, like a flower vase. The other way is to buy something used – used clothing, a used car, used furniture (some of the best bargains and most beautiful items in homes that I know are retrieved from Craigslist or eBay, or used items stores, thrift shops or antique shops. Some of the best bargains in electronic goods, such as computers, can be purchased from the manufacturer, who repairs a product and makes it like new – with a guarantee to go with it.)

Recommended Actions:

Before you throw something out, consider:

- Is there another use for it?
- Could I give it away to someone who needs it?
- Could I put it on craigslist, freecycle or ebay?
- Would it make a good toy for my child?

Before you buy something new, consider:

- Could I get a used one that would work just as well?
- Could I borrow one from a friend or neighbor?

We get a big laugh in our family when we receive a gift, and unwrap it very carefully so that the wrapping paper, ribbon and other packaging materials can be used a second time.

Children are experts at creating ideas to reuse things. Often they will trade toys, books, clothing, and other useful items. In that way each child gets something she has not had before – it's like getting a new product for nothing, and at the same time getting rid of something you are tired of. Children can often find ways to make a toy out of a paper box or paper towel roll, another example of reuse.

Adults sometimes go so far as trading homes to take a vacation, instead of paying for expensive hotel or resort fees. See www.trading-homes.com, or www.homeexchange.com, or www.ihen.com (International Home Exchange Network). A book filled with similar ideas is *Choose to Reuse: An Encyclopedia of Services, Business, Tools and Charitable Programs That Foster Reuse*, by Nikki and David Goldbeck.

Other things that can be reused are bags for shopping, old clothing for rags, old jars or drinking glasses for leftover food, etc. Use a pickle jar for storing buttons, a defective coffee cup for gathering nails and screws, and a reusable rag instead of paper towels.

For many generations in the Jewish community, we've had the concept of a *Gemach*, a common community share where Jews can give away gently used products to share with those who may need them. *Gemachs* are often specific to certain needs, such as wedding dresses or baby supplies.

Today, websites such as craigslist and freecycle are great places to find things that others are finished with, and that you can reuse. Before you buy something new, take a look at these websites to see if you can find what you want there for free.

You can also organize a resource sharing program in your own Jewish community. Canfei Nesharim (www.CanfeiNesharim.org) promotes a program called a "Fresh Exchange," a project for sharing toys within a community before Hanukkah, originally developed by a local leader (Jessica Haller) in Riverdale, New York. The idea is that many children use toys only very gently (or not at all) before they outgrow them. Instead of these toys being thrown away, they could make great gifts for others in the community. The toys are sold for low cost, with the proceeds donated to charity.

These exchanges can also be organized for costumes before Purim, for books, or for other products, and at any time of the year.

Put it into Action!

Reuse

Opportunities to Reuse

Things I will do this month:

Number Fourteen

Recycle

"Turn it over and over, for everything is in it." (Pirke Avot 5:24)

Compared to making things out of virgin resources, recycling saves resources and conserves energy, and also reduces the amount of waste going into our landfills. Many communities make recycling easy for their residents by picking up numerous items at curbside. Take advantage of your local community efforts, and recycle as much as you can!

Every home, office, synagogue and other communal institution should have recycling receptacles. Old copy-paper boxes or municipal curbside recycling cans can be used to collect recyclable materials. Check the rules for your municipality to see whether and how the items should be separated or labeled, and recycle as much as you can.

> **Recommended Actions:**
> - Find out what can be recycled in your community.
> - Recycle everything you can.
> - Encourage other friends, family members, and community organizations to recycle.
> - Buy products made from recycled materials where available.

We in the USA use over 80 billion aluminum soda cans every year.[85] It requires 95 percent less energy to make recycled aluminum cans, 60 percent less energy to make recycled paper, and 5-30% percent less energy to make recycled glass than it would from raw materials.[86] And, of course, it requires less aluminum, wood, and glass!

In 2008, Americans recycled 33.2% of our municipal solid waste. Americans recovered about 61 million tons (excluding composting) through recycling. Composting recovered 22.1 million tons of waste. Americans combusted about 32 million tons for energy recovery (about 13 percent).[87]

About 71% of office-type paper (4.3 million tons) was recycled, and about 65% of yard trimmings were recovered. Metals were recycled at a rate of almost 35 %. By recycling more than 7 million tons of metals (which includes aluminum, steel, and mixed metals), we eliminated an amount of greenhouse gas (GHG) emissions equivalent to removing more than 4.5 million cars from the road for one year.[88]

This is quite an accomplishment, but there is much more that can be done.

Find out about what is recyclable in your community, and make sure to recycle everything you can. Advocate for more things to be recycled, and encourage schools, synagogues, and local businesses to recycle. Do you recycle in your home, synagogue, JCC and other Jewish communal institutions? Do you have one can for trash, another for paper, another for plastic, etc.? Inform yourself about recycling options in your area, and make sure that you and your Jewish institutions use them. Recycling is a mitzvah! It protects God's earth, and is therefore surely the will of God.

You can learn more about the economic case for recycling at http://www.epa.gov/epawaste/conserve/rrr/rmd/rei-rw/pdf/factsheet_nat.pdf.

The web site http://earth911.com is an easy site to find information on recycling. All one has to do is enter your zip code and search by product, and you discover if and where you can recycle. It also lists curbside recycling options for your municipality. It's as easy as that!

Put it into Actio!

What's recyclable (Date: _____)
_____
_____
_____

What I will recycle from now on:
_____
_____
_____
_____

Section E: Buy Sustainably

As consumers, we have a great deal of power. We can make choices that will send messages to companies about what we want. We can refuse to buy products that cause harm to the environment, generate waste or treat their employees wrongly.

However, in order to access that power, we need to learn a lot more about where our products come from, their impact, and what choices are available to us. The web site www.goodguide.com is a very helpful tool in this regard, helping us to know what products are made in fair trade, are environmentally friendly, and are made in a sustainable mindset.

We can activate our consumer power by becoming "conscious consumers." This consciousness is the value system that Judaism tries to instill in us from birth to death.

In this section, we explore some areas where a change in what we purchase can make a real difference to our health and our environmental future.

Number Fifteen

Be a Conscious Consumer

Some years ago our son Yoni was visiting, and happened to see the toothpaste I was using. He asked me if I knew what was in the toothpaste — what were its ingredients. With my head hanging a bit, I answered that I had no idea. So, I looked at the ingredients, and could not pronounce half of the long list of chemicals with which I was brushing my teeth night after night. I was definitely NOT a conscious consumer. Thanks to our twin sons Yoni and Pesach, I hope I have become more conscious about my choices in what I buy to eat, to wear, and to use in every aspect of my life.

> **Recommended Actions:**
> - Buy only what you need.
> - Evaluate your personal and cosmetic products on the Skin Deep Cosmetic Database - http://www.ewg.org/skindeep/, and choose those which are healthiest for you and your family.
> - Learn environmental certifications, such as the Forest Stewardship Council, and companies with a preferable environmental record, and use this information to choose your products.

Many personal products, such as toothpaste, shampoo, makeup, and soaps include harsh chemicals and fragrances that can be harmful to one's health. Healthier options exist. The Skin Deep Cosmetic Database,[89] developed by the Environmental Working Group, provides information about the chemicals in your personal products and recommends those that are healthier for you and your family. You may pay a bit more for some of these products, but you could end up paying much more for health care if your current products make you sick, God forbid. You can find organic and natural hair care products increasingly at major chains, at whole-food grocers, or online. By exercising our power as a conscious consumer in mainstream stores, we are increasing the economic future of sustainable commerce.

The Internet is making it easier and easier for a conscious consumer to discover green ways to do everything. If you are eating out, for example, take a look at Local Harvest – www.localharvest.org – which keeps a directory of restaurants, farmers' markets, cooperatives, and farms that use sustainable practices and organic products. Their database covers the entire USA.

In the United States, the EPA certifies several different types of products as "environmentally preferable."
- For example, you can buy energy efficient products with the EnergyStar label (www.energystar.gov).
- Buy water-efficient appliances with EPA's WaterWise label, which meet EPA's specifications for water efficiency and performance. When you use WaterSense labeled products in your home or business, you can expect exceptional performance, savings on your water bills, and assurance that you are saving water for future generations. Learn more and find products at http://www.epa.gov/watersense/products/index.html.
- Buy environmentally friendly cleaning products, certified by EPA's "Design for the Environment" program, at http://www.epa.gov/dfe/pubs/projects/formulat/formpart.htm.

Where possible, buy wood and lumber products certified by the Forest Stewardship Council. Like an organic label for food, FSC certification assures consumers that wood products come from responsibly managed forests in which wildlife habitation and clean water are protected by ensuring that logging is done in an environmentally conscious and sustainable fashion.[90]

From the time we awaken in the morning until we go to sleep at night, we should keep our values before us, and be conscious of our important choices. This is particularly true in our purchases.

Every time we enter a store – for food, furniture, clothing, electronics, or buying tickets to go on a vacation and fly to a business meeting – we should do so with "kavanah." The Hebrew word "kavanah" means intention, or consciousness.

There is a beautiful verse in the biblical book of Proverbs (3:6): "B'khol d'rakhekha da-ay-hu;" "Know God in all that you do." As we seek to have God as preeminent in our consciousness in all that we do, so should the choices we make be intelligent, aware and mindful, so that our choices reflect the value system in our Torah and tradition.

In what ways can you become a more conscious consumer?

Put it into Action!

Current Personal Products and Better Choices:

Certifications to Look for:

Purchases I could delay:

Number Sixteen

Give Eco-Friendly Gifts and Greeting Cards

Jewish tradition has mandated gift-giving mainly on Purim, such as small food baskets to our friends, and charity to the poor. In modern times Jews have adopted the customs of the Western world and now give gifts on many more occasions, such as Hanukkah, birthdays and other life-cycle celebrations.

Recommended Actions:
- Give eco-friendly gifts.
- Wrap less.
- Send paperless greeting cards

The gift wrap industry accounts for approximately $2.6 billion a year in retail sales.[91] That amount exceeds the entre GDP (gross national product) of some countries, especially in Africa.[92]

If you must wrap your gift, use recycled wrapping paper, printed with soy based inks (rather than the usual wrapping paper made from trees and printed with petroleum-based inks and dyes). Or use the newspaper as wrapping paper – I've heard the Sunday comics work especially well.

You can also give gifts of activities or special time spent with loved ones. A friend shared with me that once she gave her dad a roll of quarters for Fathers' Day. Then they used the quarters to play pinball and video games together at a local arcade.

I often receive e-cards by email from family and friends, and do you know what? I never feel offended. Just the opposite – I feel flattered that they took the trouble and honored my values by not wasting paper unnecessarily. Three e-card web sites that I often use are www.Jewcards.com, www.cards.jhom.com, www.paperlesspost.com, and www.123Greetings.com.

Giving eco-friendly gifts is another way to protect the environment. Some beautiful items are available at Greenhome.com or Gaiam.com.

Many gifts use no resources at all, such as a donation to a worthy charity, in honor or in memory of a loved one or friend. For example, you can support the poor in Israel, or people who are served by Jewish Children and Family Services (Google to find the one in your community), and a host of other Jewish and non-Jewish charities.

> **Put it into Action!**
> **Ideas for Eco-Friendly Gifts:**
> _____
> _____
> _____
> _____
>
> **Who would appreciate them most?**
> _____
> _____
> _____
> _____

Organizations that will gladly send acknowledgments for such gifts are your synagogue, Hadassah, your local Jewish Federation, Jewish National Fund (www.jnf.org), American Jewish World Service, Red Cross, United Way, a wide range of Jewish environmental organizations, and countless others. See www.guidestar.org and www.greatnonprofits.org to verify authenticity and integrity of any charity to which you wish to contribute.

Number Seventeen

EcoTourism

Tourism is one of the largest industries in the world, and makes up a huge percentage of many countries. For example, in the Caribbean, tourism supplies 30% of the gross domestic product (GDP).[93]

> **Recommended actions:**
> - Reduce the impact of your upcoming trips and vacations.
> - Consider more local travel instead of long-distance plane trips to exotic locations.
> - When you travel, learn about the local environmental challenges and enjoy the local natural spaces.

The International Ecotourism Society (TIES) defines EcoTourism as "Responsible travel to natural areas that conserves the environment and improves the well-being of local people."[94] Principles of EcoTourism include minimizing impact, building environmental and cultural awareness, providing financial benefits for conservation and local people, and providing positive experiences for both visitors and hosts.

Eco-friendly tourism includes traveling to places that are enjoyable, but can also be educational, nature-related, and in a manner that minimizes negative effects on the environment.

Consider your plans for upcoming trips. Where are you planning to travel? Nearby vacations can be less expensive, less hassle to get to, and reduce your impact on the environment. Can the trip be made less wasteful in other ways? Can you see beautiful environmental sites while also protecting them?

Many Jews travel regularly to visit and tour in the land of Israel, and indeed, tourism is a large part of Israel's growing economy. The Israel Ministry of Foreign Affairs reports: "Although this industry contributes less than 3% to the GNP, it has a foreign currency added value of 85 percent (making it the added-value leader among the country's export industries) and employs some 80,000 persons. This industry's large potential is yet to be exploited, as it is a major factor in Israel's economic growth plan."[95] It is clear that tourism will continue to be a large portion of Israel's growing economy.

For your trips to Israel and other tourist attractions, follow the following basic guidelines for traveling green, offered by the Union of Concerned Scientists[96]:

- Do not litter
- Take away only your memories and your photographs – no "souvenirs" from the wild.
- Keep on trails
- Do not disturb wildlife or natural habitats
- Do not introduce foreign plants or animals
- Do not pollute water bodies with soap or detergents
- Do not buy things from endangered animals (such as products that contain ivory or tortoise shell)
- Do not waste water
- Turn off lights and air-conditioning when you leave your room
- Walk, or use the most environmentally-friendly methods of transportation to arrive at the sites you want to see
- Patronize hotels, airlines and tour operators that employ environmentally-friendly practices, such as energy conservation and recycling.
- Support locally owned businesses and follow local regulations.
- Since tourists like to take lots of photos, use digital camera instead of film. Solutions used to make prints often contain hazardous chemicals that require special treatment and disposal. It's also useful to avoid using disposable cameras, which creates more trash.

Organizations that can be of help in eco-friendly tourism include:
- Green Earth Travel - www.vegtravel.com
- Green Hotels Association – www.greenhotels.com
- The International Ecotourism Society – www.ecotourism.org
- Ecotourism Israel – www.ecotourism-israel.com
- JNF Travel and Tours –www.jnf.org/travel
- www.couchsurfing.org
- www.jewgether.org

Put it into Action!

Upcoming Trips:

How can I make them more eco-friendly?

Section F: Connect with Creation

You make springs gush forth in torrents;
They flow between the hills,
Giving drink to every wild animal;
The wild asses quench their thirst.
By the streams the birds of the sky have their habitation;
They sing among the foliage.
From Your lofty abode You water the mountains;
The earth is sated from the fruit of Your work.
Psalms 104:10-13

"We must obligate ourselves to meditate on creation. Try to understand both the smallest and greatest of God's creatures. Examine carefully those which are hidden from you."

Bahya ibn Pakuda, *Duties of the Heart*, 11th century, Spain[97]

Many environmental educators point to the fact that 21st century individuals have lost the sense of awe, of the Divine, and of the marvels of God's world. This, they say, is partly the cause of our sins against our planet. Our lack of respect and reverence for nature and the universe diminishes our desire to treat it more carefully and lovingly.

Richard Louv, author of *Last Child in the Woods: Saving Our Children From Nature Deficit Disorder*, tells of one child he interviewed who said, "I prefer to play indoors because that's where all the electrical outlets are." Most American adults spend almost all their time indoors, something our grandparents would be shocked to hear. The remedy, then, is to develop this sense of awe and wonder in our children, our students, and ourselves.

As a young student, the first book by Rabbi Abraham Joshua Heschel that I read was *Man Is Not Alone*. The last chapter has remained in my heart ever since. It is titled "The Pious Man." Rabbi Heschel's poetic words moved me very deeply, and intensified my feeling of amazement of the world in which we live, which God has created for us.

> We teach children how to measure, how to weigh. We fail to teach them how to revere, how to sense wonder and awe. The sense of the sublime, the sight of the inward greatness of the human soul and something which is potentially given to all men, is now a rare gift.
>
> Awe enables us to perceive in the world intimations of the divine, to sense in small things the beginning of infinite significance, to sense the ultimate in the common and the simple, to feel in the rush of the passing the stillness of the eternal.
>
> Human beings have indeed become primarily tool-making animals, and the world is now a gigantic tool box for the satisfaction of their needs.... Nature is a tool box in a world that does not point beyond itself. It is when nature is sensed as mystery and grandeur that it calls upon us to look beyond it.
>
> As civilization advances, the sense of wonder declines. Such decline is an alarming symptom of our state of mind. We will not perish for want of information; but only for want of appreciation.

From the Bible, especially the Book of Psalms, to the writings of Heschel and others, developing the sensitivity of awe and appreciation has taken a place of high priority in our religious scheme.

SECTION F: CONNECT WITH CREATION

The *gematria* (numerical equivalent) of God's name *"Elohim"* is 86. *"Hateva,"* which means nature, also equals 86. In the Middle Ages, the great sage Maimonides writes that meditating on the wonders of nature is one primary way we get to know God.[98]

> The way to come to love and fear God is by contemplating God's amazing words and creations and seeing the infinite wisdom expressed in them. This will bring one to love God and want to praise and glorify God. One will experience tremendous longing and yearning to know God's great name. In the words of David, 'My soul thirsts for Elohim, the Living Power' (Psalm 42).
>
> As one contemplates further on these things, one will immediately recoil in fear and awe realizing that one is a tiny, lowly creature standing with flimsy wisdom before the One Who has perfect knowledge. *Mishneh Torah, Yesodei Hatorah* 2:2

Rabbi Abraham Isaac Kook, first chief rabbi of pre-state Israel (d. 1935) wrote:

"Know the great reality, the richness of existence that you always encounter. Contemplate its grandeur, its beauty, its precision, its harmony. Be attached to the legions of living things who are constantly bringing forth everything beautiful."[99]

Rabbi Yosef Yitzhak Schneerson, the sixth Lubavitcher Rebbe (1880-1950) wrote in his memoirs: "One day in the summer of 1896, my father took me for a walk in the fields. The crops were ripening. A light breeze moved through the sheaves, ears of corn nodded and whispered to each other. My father said to me: "See my son - Divinity! Each

---

What can I do to increase my sense of awe?

movement of every ear of corn, and of every tuft of grass, was anticipated in the principal thought of the cosmic primordial Man."[100]

Albert Einstein said something very similar: "The most beautiful thing we can experience is the mysterious. It is the source of all true art and science. He to whom this emotion is a stranger, who can no longer pause to wonder and stand rapt in awe, is as good as dead – his eyes are closed."

If we could encapsulate these teachings, and this attitude, as individuals and as a whole in our society, we would behave much differently towards our environment.

Peter Goldberg, an attorney and a member of Congregation Shir Hadash, in Milwaukee, has created a meditation using the template of the Creation of the World as a way to focus on the daily blessings that flow to us from our radically amazing planet. See Figure 7.

Figure 7

Peter Goldberg's meditation on the how the days of creation help us focus on nature, as mentioned in Section E: Connect with Creation.

Contemplating the Days of Creation

Day One: Being, Light, Night and Day, Evening and Morning

We Jews take as our watchword the command to "hear": at Sinai God spoke and we heard. During creation, though, God "looked" at the light, the waters, the earth, the plants, the fish, the birds, the animals, at humanity and the whole of Creation and saw them as good, indeed the whole as very good.

Have you recently looked, simply looked at something, at the world?

When was the last time since childhood you lay on your back and watched the clouds scud by, or savored a whole sunset or sunrise, or watched a butterfly flit from flower to flower?

Have you allowed yourself recently not to think, but just to be?

Have you stopped to notice, to feel Buber's I-Thou relationship with anything, beyond our daily I-its?

Have your looked into yourself lately? Have you honored your sensibilities, your intuitions?

Have you honored the mysteries of the world all about you?

Have you felt Heschel's "radical amazement" at Creation?

Have you not simply passed from day to night, from morning to evening, but rather felt the rhythms of the passage from morning to evening, day to night?

Have you looked into the dark as a see-r would and allowed yourself to dream before your daily activities again start apace?

Will you look into *tohu va'vohu*, the chaos, the waste all about us and have the courage to say, "Let there be light" and you, yourself, even if all alone, to begin creation again?

Day Two: The Firmament, the Atmosphere

On a frigid winter's night have you felt the cold of space envelope the earth?

On a stormy day, have you felt and contemplated the power of a gust of wind?

Have you recently watched the fullness of tree branches swaying in response to an unseen wind? Have you run or biked against the wind, or felt the rush of the wind as you coasted downhill?

Have you heard the late summer crickets' love songs or a flute being practiced on the evening breeze? Have you breathed deeply and sighed at the fragrance of lilacs on a spring day, and do you recall the sweetness of ambrosia chocolate in the Fall air downtown?

Have you awakened to the hacking of an asthmatic child?

Have you curtailed a run or bike ride due to an ozone warning?

Have you written to a legislator about the renewal of the Clean Air Act, the destruction of forests and lakes due to acid rain, mercury in the fish due to unscrubbed coal plant emissions? Did you speak out about the new, huge WE Energies coal-powered power plant in Oak Creek [a power plant for southeastern Wisconsin], or simply write it off as an economic necessity?

Have you tried fans rather than air conditioners on moderate summer days?

Have you fought against the cut-back on public transit and rather for its extension into the suburbs and the exurbs? Have you taken public transit or biked to work? Do you drive three blocks to the convenience store instead of walking?

Have you done anything about global warming besides changing the light bulbs?

Would you be willing to pay a carbon tax on goods to pay for the social, environmental and security costs of obtaining cleaner and sustainable energies?

Day Three: Water, Earth, Plants and Diversity

Have you of late sat an hour and watched the waves of our great lake roll in or crash over a breakwater? Have you watched a fast running stream smooth then dance around a rock? Did you make it out to Kettle Moraine [glacial formations in Wisconsin] or up to the North Woods this summer? Have you ever seen the towering painted rocks of the Lake Superior shore?

Have you ever detoured from the Miller Park [stadium for the Milwaukee Brewers baseball team] parking lot to the Hank Aaron trail [Wisconsin State Park hiking and bicycling trail that runs through Milwaukee] and seen the clear depths and flowered banks of the fast running Menominee [river that runs through Milwaukee County]?

Do you recall kicking or prying apart a rotting log and watching the next layer of earth tumble out as seedbeds for future trees?

Do you garden; do you compost?

Have you *kvelled* at the abundance of tomatoes in late summer or savored the hint of mint, basil or sage in a breeze over the herb garden?

Do you concern yourself with the fields of corporate-grown corn and soybeans expanding obesely over the agricultural landscape of Wisconsin and America? Do you worry about the death of the family farm? Do you buy produce locally?

Did you decry the plumes of filthy water in Lake Michigan but support the widening of a road to cut a few minutes off your commute, thus creating acres more of run-off? Did you apply pesticides or phosphate fertilizers for a golf course quality lawn?

Have you enjoyed a burger or banana at the cost of tropical forest cut down for agricultural monocultures? Have you worried about the deforestation both north and south when you remodeled with hardwood floors and cabinets?

Did you think only of the convenience when new commercial development along Mequon and Port Roads destroyed more wetland? Have you in search of better schools or of a larger backyard been an unthinking part of urban sprawl?

Have you written your governor about his impending Great Lakes policy?

What single public health measure would most dramatically decrease world-wide mortality rates with all of the effects which flow from that?

Have you recently attended an Earth Day event?

Day 4: The Celestial Bodies, Time, Years, Seasons and Days

Did you go to see the Perseid showers this year on the very dark first night of Elul? Did you wake up early to see the eclipse of the moon? Have you taken your kids into the country or even up north to see the Aurora Borealis light up the night? How many years has it been since you got away from the light pollution of the city to marvel at the depth, solidity and brilliance of the Milky Way?

SECTION F: CONNECT WITH CREATION

Have the scientific revelations of the macro- and micro-cosmos affected your belief in a God of Creation?

Do you live by industrial and commercial time, or soccer and ballet lesson time, so that you have lost touch with natural time?

Are you aware of the cycles of the moon and the Jewish months?

Do you honor Shabbat, or is that when chores get done?

Do you know and savor the seasonal underlay of the holidays, or have they become mostly a matter of history and ritual?

Do you pause for morning, afternoon and evening prayers? What if you did?

Have you worried about the diminishing enforcement of wage and hour legislation and, absent union enforcement, the tyranny of compelled overtime or part-time work without benefits?

The name, Hebrew, is said to derive from the ancient Semitic word, *apiru*, nomadic wanderers or border-crossers. Does this effect your position on seasonal migrant workers or "illegal" immigration?

Have you thought what the convenience of 24/7 shopping might actually be costing us?

Day Five: Creatures of the Skies and the Seas, Fruitfulness, Plenitude and Sufficiency

Have you watched the birds play in the skies and dolphin play in a bay? Is this actually play, or are they all business, constantly seeking sustenance?

Have you snorkeled over a coral reef and wondered at its teeming beauty? Have you ever seen salmon fight upstream and over falls? Have you thrilled to the first geese heading north in late winter or marveled at their numbers at the Horican Marsh {major Wisconsin resting spot for millions of migratory birds] in the fall? Have you watched an eagle soar, a hawk dive, or a heron stalk and strike a fish along a Wisconsin river?

As you enjoy a salade nicoise, a sushi assortment, or a snapper dinner, do you worry about the depletion of the oceans' fisheries?

Have you ever given thought to the enormity of portions we Milwaukeeans expect at restaurants, or complained of the smaller portions at fancy restaurants? Have you ever complained that you have eaten too much? Do you worry about the continued existence, indeed inadequacy of food pantries decades after they were instituted as a temporary measure in a recession? (Don't forget to bring non-perishables for the Jewish community food pantry at Yom Kippur!)

What does species and habitat depletion really matter to you?

What will you think of during the forthcoming fast?

What is the fast that is asked of you according to Isaiah?

When is enough, enough, and what is enough?

Day Six: Wild and Domestic Animals, Wildness, Domesticity, Sufficiency and Responsibility

Thoreau said that in wildness is humanity's salvation. What about you remains wild?

Have your prayers ever sounded with or in your wild-side?

Should the "wilderness" have an innate meaningfulness for a Jew? Does it to you?

Rabbi Kook, the first chief rabbi of the Land of Israel, contended that humanity was created vegetarian, based on Gen. 1:29. What then would be the purpose of domesticated animals in the scheme of Creation?

The prefix "eco-" as in "ecology" or "economics" derives from the Greek word for household. What is the significance of the domesticated potential for Creation?

When the deer eat your tulips or a coyote, your cat, do you wonder why they are suddenly on your doorstep?

When you eat a burger or a steak, have you thought of the impact cattle-breeding has on North and South America's natural environment?

Was the scandal over the lingering deaths of cattle in the kosher slaughterhouse more than a briefly shocking read for you?

Who or what are the plural "us" in Gen. 1:26, who created humankind "in our image, in our likeness" to rule over all life forms?"

When you have read that command that humankind "master" and "rule" over nature, Gen.1: 28, have you ever wondered whether the Israelite concept of "covenant" applies to this relationship?

What of humanity's purpose in Gen. 2:15 "to tend and to till" the Garden of Eden; does that extend beyond that primal paradise to the hard-scrabble life of the world at large beyond the mythic realm?

The great Wisconsin environmentalist, Aldo Leopold, denounced the biblical notion of humanity's hegemony over nature as the source of our destructive carelessness. Do you see in Genesis 1 (and 2), the source of a Jewish concept of nature amenable to our precarious times?

Day Seven: Rest and ReSouling (*Shavat V'Yinafash*)

Was it humankind or the whole of creation which the God of Creation pronounced "very good"?

Is it humanity's world or God's world which [sic] might be holy/wholly?

German philosopher Frederick Nietzsche pronounced the "Death of God" and Romanian sociologist Mircea Eliade observed we had de-sacralized nature. Can we re-sacralize nature in a world of increasing scientism and without the death of monotheism?

Our Judaism was born of a people wandering in the wilderness. What was it (and might it still be) about the wilderness that gives birth to a sense of morality and ethics?

Do you hike or stroll on Shabbat? When you stroll, simply stroll, what do you notice?
When you slow down in any way, what do you notice?
What do you do to escape thoughts and worries of work, of daily life, of family, of finances? Does it involve nature encounters? Does it involve pets, fishing, camping, water, or gardening? What is it about such escapes that make them so moving and soulful?

If it is time with your children, have you thought of the environmental world you are leaving them? What would you like to leave them, and what are you prepared to do to secure that future?

If it is your Judaism, what is it about your beliefs and practice that so effectively move you to another plane of existence, feeling and meaning?

Do your encounters with nature, as opposed to modern urban culture, entertainment culture, technological pursuits, involve similar soulfulness?

Are you willing to sacrifice any of your material, energy-intensive and time-consuming life-style for the sake of environmental security and the future livability of the world?

Number Eighteen

Spend More Time Outdoors

A Prayer of Rebbe Nahman of Breslov

Translated by Rabbi Simkha Y. Weintraub, LCSW
Rabbinic Director, Jewish Board of Family and Children's Services

Hashem:

Grant me the ability to be alone!
May it be my custom to go outdoors each day among the trees and grass
Among all growing things,
and there may I be alone,
and enter into prayer,
to talk with the One to whom I belong.
May I express there everything in my heart,
and may all the foliage of the field
all grasses, trees and plants awake at my coming,
to send the powers of their life into the words of my prayer
so that my prayer and speech are made whole
through the life and the spirit of all growing things,
which are made as one by their transcendent Source.
May I then pour out the words of my heart
before your Presence like water, Hashem,
and lift up my hands to You in song,
on my behalf, and that of my children!

> **Recommended Actions:**
> - Find time to spend outdoors. As often as every day, if you can.
> - Learn about your local environment: parks, streams, trees, native plants, animals, as a way to connect with nature.
> - Get to know a local park or regional wild area through hiking or camping.
> - Experiment with growing or harvesting your own food.

"Nature is of the very essence of Deity."   Yisrael Baal Shem Tov, *Shivhay Ha-Besht*

There is so much to admire and to love.... Look at the sea, the sky, trees, flowers! A single tree – what a miracle it is! What a fantastic, wonderful creation this world is, with such diversity.
Pablo Casals

The world is too much with us; late and soon,
Getting and spending, we lay waste our powers;
Little we see in Nature that is ours;
We have given our hearts away, a sordid boon!
This Sea that bares her bosom to the moon,
The winds that will be howling at all hours,
And are up-gathered now like sleeping flowers,
For this, for everything, we are out of tune;
It moves us not.--Great God!
William Wordsworth, "The World is Too Much With Us"

The three pilgrimage festivals of Judaism, Pesah, Shavuot and Sukkot, began when our ancestors were primarily farmers. They are called "pilgrimage" holidays ("Shalosh Regalim") because the farmers would wend their way to the Holy City of Yerushalayim to the Bet Mikdash (Holy Temple) built by King Solomon on the Temple Mount.

There at the altar Aaron and his fellow kohanim (priests) would accept the offerings and bless the pilgrims with the famous three-fold "priestly blessing," ("Birkat Kohanim").

Only later in Jewish history, scholars tell us, did the three Pilgrimage Festivals attach themselves to an overlay of historical dimension: Pesah became a remembrance of the going out of Egypt; Shavuot, the time of the Giving of the Torah; Sukkot, the forty years of wandering, living in temporary huts.

As farmers our ancestors were familiar with the importance of the rising of the sun and its going down; with the seasons of rain and drought; with the growth of food and the need for rain (hence our prayers for "geshem" (rain), and "tal" (dew) during the Shalosh Regalim. They were outdoors folk, who marveled at the sky and its sparkling population of sun, moon and stars. They were in awe of the growth of flowers, bushes and trees to beautify God's world, and wheat, corn and rye, to nourish their bodies. "Yours is the day. Yours is the night.... You made summer and winter" (Psalms 74:16-17).

As city people and townsfolk, we have lost touch with the beauty of nature, and the marvel of its growth, its hibernation in winter and bursts of awakening in Spring.

In Rebbe Nahman's prayer above we see the need of pious teachers of the past to commune with nature. It is there that we feel God's presence most keenly, and the deep feelings of awe well up inside our heart and soul.

We need to go camping more, to ski and sled more, to hike and bike more, to tour through the great National Parks in our country and others – the Grand Canyon, Hawaiian volcanoes, Mount Rainier, Yosemite, Zion, Yellowstone, Bryce, etc. etc. etc. We need to see, in person or vicariously, The Great Barrier Reef in Australia, the Canadian Rockies, the Alps, Machu Picchu in Peru, the Egyptian Pyramids, Petra in Jordan, the Iguazu Waterfalls on the border of Argentina and Brazil or Niagara Falls in Canada, - and so many other breathtaking sites.

Our world is far more beautiful and incredible than any of us can possibly imagine. Besides the extraordinary places we just mentioned, we all have beautiful parks or rivers or mountains not far from our home where we can spend more time, and contemplate the meaning of our lives, and the beautiful gifts God gave us in our wondrous planet.

We also need to return to our historical tradition and dig in the dirt, whether in a backyard garden, a community garden, or by picking our own fruits and vegetables in the summertime. It's critical that we all remember where our food comes from, and maybe even help produce some of it. It doesn't come from the supermarket! It comes from the ground. This is an especially important project for children.

Spending time in nature helps us feel the grandeur of our planet, and brings us closer to God. The following Hasidic tale illustrates this idea:[101]

The child of a certain rabbi used to wander in the woods. At first his father let him wander, but over time he became concerned. The woods were dangerous. The father did not know what lurked there.

He decided to discuss the matter with his child. One day he took him aside and said, "You know,

I have noticed that each day you walk into the woods. I wonder why you go there." The boy said to his father, "I go there to find God."

"That is a very good thing," the father replied gently. "I am glad you are searching for God.

But, my child, don't you know that God is the same everywhere?"

"Yes," the boy answered, "but I am not."

The celebrated, greatly admired Hebrew poet of nature, Saul Tchernikovsky, expressed it in this poem,

And if you ask me of God, my God
"Where is God that in joy we may worship?"
Here on earth too God lives, not in heaven alone
A striking fir, a rich furrow, in them you will find God's likeness,
Divine image incarnate in every high mountain.
Wherever the breath of life flows, you will find God embodied.
And God's Household? All beings: the gazelle, the turtle, the shrub,
the cloud pregnant with thunder . . . God-in-Creation is God's eternal name.[102]

In her celebrated diary, Anne Frank wrote:

> The best remedy for those who are afraid, lonely, or unhappy is to go outside, somewhere where they can be quite alone with the

heavens, nature and God. Because only then does one feel that all is as it should be, and that God wishes to see people happy, amidst the simple beauty of nature. As long as this exists, and it certainly always will, I know that then there will always be comfort for every sorrow... And I firmly believe that nature brings solace in all troubles.

The Russian writer, Fyodor Dostoyevsky, wrote[103] in the spirit of ancient Judaism:

Love all God's creation, the whole and every grain of sand in it. Love every leaf, every ray of God's light. Love the animals, love the plants, love everything. If you love everything, you will perceive the divine mystery in things. Once you perceive it, you will begin to comprehend it better every day. And you will come at last to love the whole world with an all-embracing love.

In the words of Rachel Carson (1907-1964) marine biologist and author:

"If a child is to keep alive his inborn sense of wonder without any such gift from the fairies, he needs the companionship of at least one adult who can share it, rediscovering with him the joy, excitement, and mystery of the world we live in."[104]

Rabbi Lawerence Kushner, in his book, *Honey From The Rock,* writes: "To be a Jew means to wake up and to keep your eyes open to the many beautiful, mysterious, and holy things that happen all around us every day."

Use these thoughts to inspire you to spend more time outdoors and to appreciate creation.

Put it into Action!

Places I find inspiring:

Camping/Hiking trips I could plan:

Times I will spend outside:

Number Nineteen

Learn and Teach the Jewish Prayers of Thanksgiving for the World

Praised are You, Our God, Ruler of the universe, former of light, creator of darkness, maker of peace and the creator of all things. In Your mercy light shines over the earth and upon all who inhabit it. Through Your goodness the work of the creation is daily renewed. How great are Your works, O God, in wisdom You have made all of them. The earth is filled with your creations.
Daily Prayer Book, Yotzer Or

> **Recommended Actions:**
> - Learn and recite Jewish blessings of appreciation.
> - Use this as an opportunity to become more conscious of the blessings of the world and in your life

Could song fill our mouth as water fills the sea
And could joy flood our tongue like countless waves,
Could our lips utter praise as limitless as the sky
And could our eyes match the splendor of the sun,
Could we soar with arms like eagle's wings
And run with gentle grace, as the swiftest deer,
Never could we fully state our gratitude
For one ten-thousandth of the lasting love
Which is Your precious blessing, dearest God,
Granted to our ancestors and to us.
Shabbat and Festival Prayer Book – "Nishmat Kol Hai"

The Talmud teaches (Berakhot 35a):

Anyone who enjoys the natural world without a blessing, it is as if he has enjoyed from that which is sanctified for heaven. As it is written in Psalms: "The earth is the Lord's and all that it holds" (Psalms 24:1). Rabbi Levi pointed out that this verse is contradicted by another verse in Psalms that says "The heavens are God's Heavens, but the earth was given to humankind" (Psalms 115:16). However, explains the Talmud, this is not a contradiction. The former verse is true *before* one says a blessing, and the latter verse is true *after* one says a blessing.

In another tractate of the Talmud, Rabbi Meir said: A person is obligated to say one hundred blessings every day. Menahot 43b

The Jewish liturgical tradition includes numerous blessings, which were designed to heighten human sensitivity to the everyday blessings of life. Learning and reciting these blessings can help us slow down and become more grateful for the blessings of the natural world. Give this a try for a day or a week, and see the difference it makes for you to be present to the wonders of creation and your opportunity to appreciate them. For example:

Blessings of Gratitude

The insights of wonder must be constantly kept alive. Since there is a need for daily wonder, there is a need for daily worship.... This is one of the goals of the Jewish way of living: to experience commonplace deeds as spiritual adventures, to feel the hidden love and wisdom in all things.

-- Rabbi Abraham Joshua Heschel

Prayer over the washing of the hands in preparation for the b'rakah over bread-

Praised are you Ado-nai our God, Who rules the universe, instilling in us the holiness of the mitzvot by commanding us to rinse our hands.

Before eating bread-

Praised are you Ado-nai our God, Who rules the universe, bringing forth bread from the earth.

Before eating foods made of the five grains - wheat, barley, oats, rye, or spelt-

Praised are you Ado-nai our God, Who rules the universe, creating various grains.

Before eating fruit-

Praised are you Ado-nai our God, Who rules the universe, creating the fruit of the tree.

Before eating vegetables (from plants which sprout annually)

Praised are you Ado-nai our God, Who rules the universe, creating the fruit of the earth.

Before drinking wine or grape juice-

Praised are you Ado-nai our God, Who rules the universe, creating the fruit of the vine.

Before partaking of other foods or liquids-

Praised are you Ado-nai our God, Who rules the universe, through whose world all things exist.

When eating a new food, or a seasonal fruit not tasted since the previous season. add this b'rakhah before eating:

Praised are you Ado-nai our God, Who rules the universe, granting us life, sustaining us, and enabling us to reach this day.

Blessings of Appreciation for Nature:

Upon seeing the wonders of nature- lightning, shooting stars. vast deserts, high mountains. a spectacular sunrise or sunset-

Praised are you Ado-nai our God, Who rules the universe, renewing the work of creation.

Upon seeing a storm or hearing thunder-

Praised are you Ado-nai our God, Who rules the universe, Whose power and might fill the universe.

Upon seeing a rainbow-

Praised are you Ado-nai our God, Who rules the universe, faithfully recalling the covenant by keeping the divine promise.

Upon seeing the ocean-

Praised are you Ado-nai our God, Who rules the universe, having fashioned the great sea.

Upon seeing trees in bloom for the first time each year-

Praised are you Ado-nai our God, Who rules the universe, which lacks nothing; for God created fine creatures and pleasant trees in order that humans might enjoy them.

Upon seeing creatures or vegetation of striking beauty-

Praised are you Ado-nai our God, Who rules the universe, in Whose world such beauty exists.

Upon seeing unusual creatures-

Praised are you Ado-nai our God, Who rules the universe, diversifying creation.

Adapted from *Siddur Sim Shalom for Weekdays.* The Rabbinical Assembly

> **Put it into Action!**
> **Which blessings would I like to say regularly?**
> _____
> _____
> _____
> _____
>
> **How will I remind myself to do this?**
> _____
> _____
> _____
> _____

What other biblical texts, psalms, and prayers in the Siddur do you know that enhance our appreciation of nature and our universe?

Number Twenty

Observe Shabbat

Rabbi Samson Raphael Hirsch wrote:

Sabbath in our time! To cease for a whole day from all business, from all work, in the frenzied hurry-scurry of our time? To close the exchanges, the workshops and factories, to stop all railway services –

> **Recommended action:**
> - Slow down one time a week on the Sabbath, and enjoy a day of rest. The earth will enjoy it too.

great heavens! How would it be possible? The pulse of life would stop beating and the world perish!' The world perish? On the contrary – it would be saved. (Samson Raphael Hirsch, 19th Century German Orthodox Rabbi, *Judaism Eternal*)[105]

Shabbat has many meanings for many people.[106] Shabbat can play an important role in developing a sense of awe, by pausing and slowing down. This can have a real value for protecting the environment.

What is the essence of Shabbat? Shabbat is a day of rest, not only for humans, but also for animals, and for the land.

The Torah mentions the Shabbat in several places. God created the earth and all that is in it for six days and rested on the seventh (Gen. 2:1-3). In the Ten Commandments (Exodus 20) we are instructed to "Remember the Sabbath day and keep it holy... you, your son or daughter, ..."

The concept of Shabbat is also referenced in Leviticus 25:4, where we are directed that "in the seventh year there shall be a Sabbath of solemn rest for the land, a Sabbath to Ado-nai. You shall not sow your field or prune your vineyard." The Shmitah year, every seventh year, is another time when work is prohibited for an entire year, so that the land can rest. This has positive ecological benefits for the soil.

Rabbi Norman Lamm, chancellor of Yeshiva University, writes:

> Perhaps the most powerful expression of the Bible's concern for man's respect for the integrity of nature as the possession of its Creator, rather than his own preserve, is the Sabbath.... The six workdays were given to man in which to carry out the commission to "subdue" the world, to impose on nature his creative talents. But the seventh day is a Sabbath; man must cease his creative interference in the natural order (the Halakha's definition of melakha or work), and by this act of renunciation demonstrate his awareness that the earth is the Lord's and that man therefore bears a moral responsibility to give an accounting to its Owner for how he has disposed of it during the days he "subdued" it.... A new insight into Jewish eschatology: not a progressively growing technology and rising GNP, but a peaceful and mutually respectful coexistence between man and his environment.[107]

What the Torah stresses in its prohibition of performing any creative work on the Shabbat is that in six days we are to work the soil, create new things, develop technology, work the land and its produce in every way possible, but on the seventh day we must not tamper with God's world.

The Shabbat is thus a testimony to the importance of respecting God's creation, our universe and our planet, by not interfering with it in any way. Psychoanalyst Erich Fromm, in his book, *The Forgotten Language*, points out that the distinguishing characteristic of the Jewish Sabbath is the prohibition of doing anything to change or alter nature. Nothing creative – no building, no planting, no altering God's world, is permitted – manifesting deep respect for the world as God created it.

Rabbi Samson Raphael Hirsch (1808-1888), champion of neo-Orthodoxy in Frankfurt, Germany, saw the importance of the Sabbath in a similar way. To Rabbi Hirsch, the Sabbath was commanded to the Jewish People, "in order that they should not grow overweening in their dominion" of God's creation...." Jews "should refrain on this day from exercising their human sway over the things of the earth, should not place a hand upon any object for the purpose of human dominion, that is, to employ it for any human end; we must, as it were, return the borrowed world to its Divine Owner in order to realize that it is but lent to us. On Shabbat we strip ourselves of our glorious mastery over the matter of the world, and lay ourselves and our world acknowledgingly at the feet of the Eternal our God."[108]

Those who are committed to preserving our planet and sustaining it for future generations are concerned that by slavishly adhering to a seven-day work week, we do not give ourselves either the time or the attitude to consider what our efforts are doing to harm the planet.

Imagine the opportunity of Sabbath in our time. No driving, no internet, no television, no commerce. The opportunity to rest and be with family. More than that, infusing a day with sacredness through the lighting of candles, reciting of Kiddush, and some prayer, makes every day feel a bit more holy.

By resting one day a week, and letting the land rest, we are acknowledging that we are neither the owners nor the masters of the earth. That our Creator put us here to "work the soil and also to preserve it" (Genesis 2:15). We do a great job of working it, but have not been as careful as we should in preserving it.

Put it into Action!

How could a "day of rest" make a difference for me and my family?

_____
_____
_____

What Sabbath actions (or non-actions) could I experiment with?

_____
_____
_____

Section G: Greening Your Home

"What is the use of a House if you haven't got a habitable Planet to put it on."

Henry David Thoreau

The home is the center of Jewish life. In Jewish tradition, it is known as "mikdash m'at," or "a small sanctuary."

We spend a great deal of time at home, as Jews should, especially in evenings and weekends. A traditional Jew who observes Friday night at home with all the beautiful rituals will spend most of Friday night with the family, and much of Shabbat afternoon as well.

We can make our home an even holier place if we become more conscious of the ways in which we help preserve our environment in the home (use of air, water, cleaning materials, etc., as we shall see as we explore these areas).

You can explore potential environment and health dangers in your home on an interactive website, a collaboration of WebMD and Healthy Child Healthy World: http://www.webmd.com/health-ehome-9/default.htm

For a complete list of ways to "green your home" check out the U.S. government's web site for the Environmental Protection Agency (EPA) - www.epa.gov, which includes topics such as:

Protect the Environment: At Home and in the Garden

http://www.epa.gov/epahome/home.htm,

Saving Energy

Find Energy Star products for your home –

http://www.energystar.gov/index.cfm?fuseaction=find_a_product.

Choosing energy-efficient products can save families about 30% ($400 a year) while reducing our emissions of greenhouse gases. Whether you are looking to replace old appliances, remodel, or buy a new house, you can help. ENERGY STAR is the government-backed symbol for energy efficiency. The ENERGY STAR label makes it easy to know which products to buy without sacrificing features, style or comfort that today's consumers expect.

More steps you can take:

- Turn off appliances and lights when you leave the room.
- Use the microwave to cook small meals. (It uses less power than an oven.)
- Purchase "green power" for your home's electricity. (Contact your power supplier to see where and if it is available).
- Have leaky air conditioning and refrigeration systems repaired.
- Cut back on air conditioning and heating use if you can.
- Insulate your home, water heater and pipes.

Read these sections on the EPA web site: www.epa.gov

Reducing Air Pollution and Greenhouse Gas Emissions

Climate change: what you can do at home and in the garden –

http://www.epa.gov/climatechange/wycd/home.html

SECTION G: GREENING YOUR HOME

Making a few small changes in your home and yard can lead to big reductions of greenhouse gas emissions and save money.

Conserving Water

Choose water-efficient products and test your WaterSense —

http://www.epa.gov/watersense/water_efficiency/cons.html

More steps you can take:

- Don't let the water run while shaving or brushing teeth.
- Take short showers instead of tub baths.
- Keep drinking water in the refrigerator instead of letting the faucet run until the water is cool.
- Scrape, rather than rinse, dishes before loading into the dishwasher; wash only full loads.
- Wash only full loads of laundry or use the appropriate water level or load size selection on the washing machine.
- Buy high-efficient plumbing fixtures & appliances.
- Repair all leaks (a leaky toilet can waste 200 gallons a day).
- Water the lawn or garden during the coolest part of the day (early morning is best).
- Water plants differently according to what they need. Check with your local extension service or nurseries for advice.
- Set sprinklers to water the lawn or garden only - not the street or sidewalk.
- Use soaker hoses or trickle irrigation systems for trees and shrubs.
- Keep your yard healthy - dethatch, use mulch, etc.
- Sweep outside instead of using a hose.
- Landscape using "rain garden" techniques to save water and reduce stormwater runoff. Video: "Reduce Runoff: Slow It Down, Spread It Out, Soak It In" - http://owpubauthor.epa.gov/polwaste/green/video.cfm

Reducing, Reusing and Recycling Materials

Practice the three R's: first **reduce** how much you use, then **reuse** what you can, and then **recycle** the rest. Then, dispose of what's left in the most environmentally friendly way. Read the tips below and explore the <u>Consumer's Handbook for Reducing Solid Waste</u>,

http://www.epa.gov/epawaste/wycd/catbook/index.htm

- Reduce:
    - Buy permanent items instead of disposables.
    - Buy and use only what you need.
    - Buy products with less packaging.
    - Buy products that use less toxic chemicals.
- Reuse:
    - Repair items as much as possible.
    - Use durable coffee mugs.
    - Use cloth napkins or towels.
    - Clean out juice bottles and use them for water.
    - Use empty jars to hold leftover food.
    - Reuse boxes.
    - Purchase refillable pens and pencils.
    - Participate in a paint collection and reuse program.
    - Donate extras to people you know or to charity instead of throwing them away.
- Recycle:
    - Recycle paper (printer paper, newspapers, mail, etc.), plastic, glass bottles, cardboard, and aluminum cans. If your community doesn't collect at the curb, take them to a collection center.
    - Recycle electronics. http://www.epa.gov/epawaste/conserve/materials/ecycling/index.htm
    - Recycle used motor oil. http://www.epa.gov/osw/conserve/materials/usedoil/index.htm
    - Compost food scraps, grass and other yard clippings, and dead plants.

SECTION G: GREENING YOUR HOME 117

> o   Close the loop - <u>buy recycled products and products that use recycled packaging</u>. That's what makes recycling economically possible. http://www.epa.gov/epawaste/conserve/rrr/buyrecycled.htm

You can find an abundance of useful information on the government's web site – (EPA – Environmental Protection Agency - <u>www.epa.gov</u>. Below are some of the topics you can search for, including many publications offered by the EPA.

Ensuring Safe Drinking Water

Improving Indoor Air Quality

Using Toxics and Pesticides Safely

Reducing Your Exposure to Harmful Substances

Using Safer Cleaning Products

Pollution Prevention

Buying and Maintaining an Environmentally Friendly House

Lawn and Garden Care

Number Twenty-One

Clean the Green Way

"One must endeavor to keep far from a foul smell." Talmud, Tractate Horiyot 13b

There's an old Jewish joke about a housewife who cleans the house before the maid comes, lest the cleaning person see a dirty home. The comedian Rita Rudner once said: "Neurotics build castles in the air, psychotics live in them. My mother cleans them."

> **Recommended Actions:**
> - Stop using air fresheners.
> - Reduce your use of chemical cleaners, and consider more healthy options, for example products certified by EPA's Design for the Environment (DfE) program.

Jews have been known for their cleanliness. Many scholars argue that the laws of kashrut, and the laws of purity and impurity ("tum'ah and taharah"), so prevalent in the Torah (especially in the book of Leviticus) have ancient origins in health and cleanliness. The Torah demands that we "take utmost care and watch ourselves scrupulously" (Deut. 4:9), which has been interpreted by commentators throughout the ages to mean that one must guard one's health.

Cleanliness is next to Godliness, right? But <u>how</u> we make our homes clean can also bring Godliness into the home.

When you use a cleaning product, it is released into the environment: inside your home and down the drain to the outdoors. Adults come in contact with cleaning products on a regular basis, as do children who are often the most exposed when they crawl on the floor.

It's common to clean our homes with dangerous substances that come in pesticides, air fresheners, disinfectants, and other products. These hazardous materials that we breathe or stick to surfaces that we touch, hang in the air and then often end up in our bodies. A significant part of the rise in the incidence of cancer in the past century stems from the increase of chemicals in our cleaning process.[109] Cleaning chemicals can also have harmful effects on the reproductive, immune and nervous systems.[110]

These harmful chemicals not only hurt our bodies, they also harm the environment. The example of Los Angeles comes to mind, where the heavy smog a visitor sees (and chokes on) is not only the result of tailpipe emissions, but also, perhaps even more so, from household products.[111]

Further, these hazardous chemicals that we use to clean the toilet bowl, kitchen sinks and bathroom fixtures, are often flushed down the drain and then enter our water system. Experts in this field tell us that more than 200 chemicals commonly found in pesticides, cosmetics, dyes, drugs and gasoline and diesel exhaust can cause breast cancer.[112]

Avoid cleaning products with lye or acid, such as drain cleaners, oven cleaners and some toilet bowl cleaners. Many cleaning products also contain "Volatile Organic Compounds" (VOCs) which have the ability to irritate and harm lungs and respiratory systems. Chlorine bleach and ammonia are dangerous on contact, and can enter ecosystems when they become airborne from wastewater.

The federal government does not currently test air fresheners for safety or require manufacturers to meet any specific safety standards. A 2007 study of the Natural Resources Defense Council (NRDC)[113] found that most contain chemicals that may affect hormones and reproductive development, particularly in babies.

EPA's "Design for the Environment Safer Product Labeling Program" uses EPA's chemical expertise and resources to evaluate products carefully and to label only those that have met the program's highly protective standards. The DfE logo is an easy way to know you are choosing a product that is as safe as possible for people and the environment.

EPA's Design for the Environment Program (DfE) has allowed use of their logo on over 2000 products. DfE labels a variety of chemical-based products, like all-purpose cleaners, laundry detergents, and carpet and floor care products. These products are formulated from the safest possible ingredients and have reduced the use of "chemicals of concern" by hundreds of millions of pounds. Products in the DfE program do not sacrifice quality or performance —and are safer for people and the planet.

Look for the Design for the Environment logo when you shop. To find cleaning products that have been approved by EPA's Design for the Environment program, visit http://www.epa.gov/dfe/pubs/projects/formulat/formpart.htm.

More interested in making your own? Here are some recipes for homemade eco-friendly cleaning products, courtesy of Canfei Nesharim (www.CanfeiNesharim.org):

1. Glass Cleaner (for windows and mirrors)
Fill 16 oz. spray bottle with ¼ cup white vinegar and 1/4 tsp liquid detergent. Fill to the top with water. Shake to combine.

2. Wood Floor Cleaner
Mix 1/4 cup white vinegar/lemon juice and 1/4 cup olive/veg. oil in a bottle.

3. Furniture Polish
In a spray bottle mix 1/2 tsp veg. or olive oil and 1/4 cup white vinegar or lemon juice. Shake well and apply small amount to cloth.

4. Bath and Tile Cleaner
Make a soft scrub by mixing 2 cups baking soda, 1/2 cup liquid castille soap, 4 tbls vegetable glycerin and a couple drops of vinegar. Apply, scrub and wipe.

{Ideas gathered from nationalgeographic.com and womenandenvironment.org. Thanks to Rena Dubensky for providing this information.}

While we're on the topic of cleaning, look for a green dry cleaner. Regular cleaners use highly toxic solvents that harm those who work there, as well as the wearers of the clothing. Further, the solvents pollute the air and water systems. If possible, avoid buying clothing that requires dry cleaning.

If our home is truly to be a "mikdash m'at," we want it to be a place filled with love, care, compassion and support. We want a home that will promote good health and a clean environment, in which we and our families are free to grow without danger of illness or harm. This is in fulfillment of Deuteronomy 4:9,15, "Guard yourself and guard yourself scrupulously," which the tradition interprets to mean that one must guard one's body and soul.

Put it into Action!

Cleaning Products to Check:

Ways I will "clean more green":

Figure 9

Top 10 Eco-Friendly Ways to Clean the House

Reprinted from Care2.com, see http://www.care2.com/greenliving/clean-house-top-10-eco-friendly-ways.html

Choosing environmentally friendly cleaning products — and removing toxic ones — goes a long way towards ensuring a home with fresh, clean air. Clean air renews and rejuvenates; it doesn't pollute our lives or the environment. Living in a less toxic home, removed from neurotoxic chemicals, improves sleep and concentration, makes babies less fussy, and gives a sense of well-being. Your household's toxic burden on the environment will be significantly reduced by following these steps, and this too can bring peace of mind.

1. LOOK UNDER YOUR KITCHEN SINK: Remove toxic products
WHY: Almost everyone in the world has a cupboard full of poisons under their kitchen sink. Wasp spray, oven cleaner, waxes and polishes—the place is full of chemicals that display the words *poison, danger, warning,* or *caution*. Small amounts of the poisons drift from, and leak out of bottles and spray bottles, which then waft around the kitchen. Household poisonings are one of the highest threats to the health of children.

HOW: Place products with signal words in a locked cupboard in storage for your community's next Household Hazardous Pickup Day (see next tip); replace all hazardous products with safer versions in the future.

HIGHLIGHTS: No chemicals wafting into your household; safer environment for kids.

SECTION G: GREENING YOUR HOME

2. ABOUT HOUSEHOLD HAZARDOUS WASTE PICKUPS: Take toxic products WHY: Hazardous materials shouldn't be poured down the drain or thrown away in the trash as they can cause serious pollution problems in the waste stream.

HOW: Call your local recycling center, town or city hall. Most communities have at least one Household Hazardous Waste Pickup Day a year.

HIGHLIGHTS: There will be fewer toxic materials leaching out of landfills, burning in incinerators, and being washed into the waste water stream.

3. REPLACE TOXIC PRODUCTS: Choose non-toxic, biodegradable substitutes WHY: Help reduce the toxic burden of manufacturing, your home, and the waste stream. HOW: Read "Signal Words" on labels. The signal words *poison, danger, warning,* or *caution,* found on the label of products such as pesticides and cleaning products, are placed there by order of the federal government and are primarily for your production. In some cases these signal words are on the label because of the potential impact the product can have on the environment. *Poison/danger* denotes a product of most concern, one that is highly toxic, and ingesting small amounts—in some cases a few drops—can be fatal. *Warning* means moderately toxic, as little as a teaspoonful can be fatal; and *caution* denotes a product that is less toxic, one in which it would be necessary to ingest between two tablespoons and two cups to be fatal. *Corrosive* products can damage skin and mucous membranes, and a *strong sensitizer* is a chemical that can increase allergies.

HIGHLIGHTS: Labels provide information by which you can protect yourself, your family, and the environment.

4. LEARN NON-TOXIC CLEANING BASICS: How to use kitchen cupboard ingredients

WHY: Save money, protect your health, reduce your use of valuable resources of the earth, avoid petroleum products and other non-renewable resources.

HOW: Learning to clean from scratch—making homemade recipes—can truly work if you take time to understand a bit about the chemistry behind how the materials work. Here are the five ingredients that Annie (the author of *Clean & Green* among other books, and Care2's Healthy Living channel producer) finds to be the safest, most effective, and useful for cleaning.

The Five Basics for Non-Toxic Cleaning

http://node1-www.care2.com/greenliving2/wp-admin/%20/greenliving/five-basics-for-nontoxic-cleaning.html

Make sure to keep all homemade formulas well-labeled, and out of the reach of children.

Note how to safely reduce four airborne allergens in the home with these simple steps.

http://www.care2.com/greenliving/control-four-airborne-allergens.html

HIGHLIGHTS: Establish a safe, cheap and simple lifestyle.

MORE: Visit Care2.com's Healthy Home category for many non-toxic cleaning tips. Just scroll down to Non-Toxic Cleaning. http://www.care2.com/channels/lifestyle/home

5. OF MOPS, SPONGES, RAGS, AND OTHER ACCESSORIES: Natural, reusable

WHY: Reduce your use of non-renewable resources; avoid products with potentially harmful ingredients such as sponges with antibacterial ingredients; reuse old shirts as rags and more. Use cloth rags instead of paper towels to save trees. Save money!

HOW: Look at your purchase of mops, paper towels, sponges, buckets, vacuums, and more with an eye towards their durability, health and environmental impact. If you must use paper towels buy recycled, unbleached paper.

SECTION G: GREENING YOUR HOME

HIGHLIGHTS: Reusable mops, rags instead of paper, safe sponges, HEPA vacuums all work towards providing your home and environment with fresh, clean air, and reduce your consumption of nonrenewable resources.

6. LEARN ABOUT YOUR WATER: Is it hard or soft? WHY: With hard water you will most likely need to clean with a detergent instead of a soap to avoid soap scum.

HOW: Read here about when and why to choose a detergent or a soap:

http://node1-www.care2.com/greenliving2/wp-admin/%20/greenliving/detergent-or-soap.html

HIGHLIGHTS: Choosing the right product for the right job reduces time and resources.

7. DISINFECTANTS? CHLORINE BLEACH? Look for alternatives WHY: Just as antibiotics are causing drug resistance, so too are disinfectants. Chlorine bleach can cause cancer causing chemicals to form in the waste water stream:

http://www.care2.com/greenliving/chlorine-in-household-cleaners.html

HOW: Make a safer antibacterial spray: (http://www.care2.com/greenliving/antibacterial-spray.html) by using these suggestions. Read here about toxic sponges: http://www.care2.com/greenliving/antibacterial-spray.html

Visit your natural food store and ask for their recommended chlorine beach alternative. Seventh Generation and other brands offer alternatives that work.

HIGHLIGHTS: A healthier home and healthier environment.

8. CONSERVE WATER

WHY: Clean water is one of our most precious and diminishing resources and we don't want to waste it. HOW: Don't run the water unless you are using it or catching it in a bucket for use; sweep instead of wet mop when possible; put a tracking matt at the door to collect mud and dust so you will need to wash the floor less; etc. Use common sense.

HIGHLIGHTS: Do your part to preserve the earth's precious resources.

9. CLEAN INDOOR AIR WITH PLANTS

WHY: Plants have been found to reduce indoor air pollution!

HOW: Here are the top 10 plants that clear indoor air:

http://www.care2.com/greenliving/antibacterial-spray.html

HIGHLIGHTS: Plants clean the air and provide more oxygen too!

10. USE YOUR SENSES: Smell, feel, hear

WHY: If you use your nose you will know when food is rotten, when dog beds need to be cleaned, when toxic chemicals may be leaking from old product bottles, and more. If you use your sense of touch you will know when doorknobs are sticky, the floor needs washing, etc. If you allow your senses to be your guide you will stay on top of cleaning jobs that need to be attended to.

HOW: Listen to what your senses are telling you.

HIGHLIGHTS: Cleaner indoor air, alert to potential toxic exposures.

Number Twenty-Two

Use Your Bathroom Wisely

One might ask a simple question: What on earth does Judaism have to say about the bathroom? In fact, there is a special blessing one is required by Halakhah to recite in Hebrew upon exiting the bathroom. It is generally known as "asher yatzer," and this is the text:

> **Recommended Actions:**
> - Save water and energy in the bathroom.
> - Take shorter showers.
> - Use eco-friendly cosmetics, deodorants, and personal care products.

> Rabbi Abayei said, when one comes out of a privy one should say: Blessed is God who has formed us in wisdom and created in us many orifices and many cavities. It is obvious and known before Your throne of glory that if one of them were to be ruptured or one of them blocked, it would be impossible for a person to survive and stand before You. Blessed are You Who heals all flesh and does wonders. (T.B. Berakhot 60b)

This is just one more example of how there is no area of life, spiritual or physical, that is not covered by some Jewish law, custom, value or interest. This is part of the comprehensive beauty of the Jewish tradition.

The bathroom is a place where many important as well as wasteful (double-entendre) things take place. We have mentioned some of them before, but it's a good idea to bring together all the ideas for this important place in our home.

Be careful with use of water in the bathroom. Water is a precious commodity. For example:
- Take a shower instead of a bath, and keep your shower short. A bath uses about 20 gallons of water and a shower only half of that (if it's

not too long). Turn the water off while you are sudsing your hair or soaping parts of your body.
- Turn the water off in the sink when you are brushing your teeth.
- Have your toilet set to use a minimum amount of water when flushing. Newer toilets do that automatically. If you have an older one, you may think of replacing it.
- If not, placing a 1.5 liter-jar filled with water in the toilet's reservoir will reduce your toilet's usage per flush. (Note: if you are using a plastic container, fill it with some pebbles in addition to the water, to weigh it down. Otherwise, your toilet may keep running, wasting even more water!)
- In European and Middle Eastern countries, it is normal to use dual-flush toilets. The flush mechanisms are separated so that one handle or button is used for small amounts to flush, and another one for larger flushes.
- Water that has been used in sinks, bathtubs and showers ("graywater") can be re-used to flush the toilet, water the lawn (if it contains only biodegradable soaps).
- New houses sometimes use graywater recycling systems. See this web site:
http://www.thenaturalhome.com/greywater.html
- When washing your hands, brushing your teeth or showering, don't put the faucet on full force. Use only what you need.
- When traveling, you'll find that many hotels give you a choice of washing your towels every day, or hanging them back up and using them more than once. Do as you do at home – and don't abuse the water on vacation just because you're not paying for it.
- When shaving, keep some water in the sink to rinse the razor, instead of rinsing under running water. You'll use far less water, and it will clean your razor just as well.
- Beware of toilet leaks! They can waste up to 200 gallons of water a day.[114] One way to know if there's a leak is to place a bit of food coloring into the tank. If the colored water shows up in the bowl, your toilet is leaking.

While in the bathroom, check the safety of your soaps, toothpaste, anti-perspirant/deodorant, and other cleansing products on the Environmental

Working Group's Skin Deep Cosmetic Database. www.cosmeticdatabase.com .

You can also install sensor lights in all bathrooms. By doing so you will save much energy. (But note that, from the standpoint of Jewish law, sensor lights still count as "turning on the lights" on the Sabbath.)

The U.S. Green Council's Green Building Guide offers a wide range of tips on greening your bathroom, from water and energy savings to eco-friendly counters and tiles. You can explore the full information at: http://greenhomeguide.com/know-how/article/making-your-bathroom-healthy-efficient-and-comfortable

Put it into Action!

How can I green my bathroom habits?

Water Savings:

Healthier cosmetics:

Number Twenty-Three

Compost

"The Lord God formed man from the dust of the earth." (Genesis 2:7)

"And the Lord God formed out of the earth all the wild beasts and all the birds of the sky...." (Genesis 2:19)

"For dust you are and to dust you shall return." (Genesis 3:19)

Every day you generate food wastes, and if you have a yard or garden, each year you collect many tons of leaf and yard debris. If placed into a landfill, these wastes will take many decades to decompose and may never return to the natural system that created them. If you can let them decompose naturally, however, you'll create a great source of nutrients to enrich your garden and local area. What is the magic that turns these wastes into fertile garden soil? Composting.

Recommended Actions:
- Consider composting your yard and food waste.
- Learn about and take advantage of your local composting resources.

When you "compost," you are allowing the remains of plants and other plant-based materials (such as fruit and vegetable peels) to turn back into the soil that created them. This is a natural process that is interrupted by putting our waste into plastic bags and then dumping them into landfills. For example, in the forest, when leaves pile up on the forest floor, they decay naturally and ultimately become the soil that fertilizes new trees.

Today, composting allows you to create your own source of fertilized soil from your yard and (as appropriate) kitchen waste. There are many different types of products that can help you, and websites which describe the process. The simplest version of composting is simply to pile your leaves and branches in a corner of your yard and let them decompose naturally. More sophisticated compost bins are also available.

Composting has two basic environmental benefits. First, you save landfill space by putting your yard and food waste back into the ground instead of the trash. Second, you create a natural fertilizer that removes or reduces the need for chemical fertilizer in your garden. (Chemical fertilizers can do significant damage to your local water sources.)

Before you begin to compost, you should check to see if your local municipality collects any items for composting. If so, they will simplify the process for you enormously by taking your waste to a large composting facility. Some municipalities collect a wide range of composting materials, some collect yard waste regularly or at high-debris times of year (such as in the fall when there are many leaves), and some do not collect at all. Some municipalities even offer free composting bins and workshops to show you how to compost. If you'd like to see more composting resources in your community, talk to your local representative.

Composting has gotten a bad rap in the past because compost bins can get smelly or become a breeding ground for bugs. A well-balanced compost bin will not present these problems, but because it can take some practice to get it right, it is important to assess your own comfort (and those in your family!) with smells or bugs before you begin.

If you or other family members are especially sensitive to these issues, you may want to consider starting very simply with yard waste in your backyard. Yard waste is unlikely to generate significant smells and the beneficial bugs it gathers will stay generally contained within the compost area you've created. You can also get a very basic compost bin for the backyard that is open on the top and bottom but will keep the yard waste set off from the rest of the yard and generally contained.

I must confess that composting is still on my list of things to do in the future. I am hopeful that I will be able to do it sooner rather than later.

If you want to compost food waste, it's probably best to buy a more enclosed compost bin. Composting food wastes outdoors in open bins can sometimes generate unwanted pests. Enclosed compost bins can be stored inside or outside, depending on the circumstances. A wide range of compost bins are available, including some which involve worms (which will eat your waste and help it decompose), turners to help the waste decompose more quickly, and even deodorizers to remove any odor.

Especially if you are composting food waste, you'll want to make sure to balance your "carbon" and "nitrogen" ratio in order to make composting work effectively. A balanced carbon and nitrogen ratio, along with proper drainage, will also help avoid any odor problems.

The EPA offers a very useful site about composting, including basic information, which materials should and should not be composted, regional and state composting programs, environmental benefits, and information for further study. The EPA website can be viewed at http://www.epa.gov/epawaste/conserve/rrr/composting/index.htm.

Some additional helpful resources can be found at:
http://vegweb.com/composting/
http://www.howtocompost.org/
http://www.composting101.com/

Put it into Action!

What I think about composting:
_____
_____
_____

Local Resources:
_____
_____
_____

Actions taken:
_____
_____
_____

Section H: Sustainability and Jewish Eating

Judaism has always laid great stress on patterns of eating, as a way to elevate the spiritual nature of our physical selves. The laws of Kashrut, for example, include a range of prohibitions that can help us develop awe and respect for living beings. The methods of kosher slaughter ("shehitah") encourage humane treatment of animals, which are killed with one sharp slit of the jugular vein in order to reduce the possibility of causing pain to an animal killed by a ritual slaughterer ("shohet"). The separation of milk and meat stems originally from the biblical prohibition of boiling a baby goat in the milk of its own mother – another way to encourage compassion to animals.

The table in a Jewish home is compared to the altar in the ancient Temple in Jerusalem. The ancient requirement to recite blessings before and after meals, when practiced, goes a long way toward raising the level of a simple animal need of providing the body with fuel to the higher plane of a divine act.

Perhaps Judaism can contribute something to the environmental world with our traditional wisdom about eating.

The great philosopher and physician Maimonides (12th century, Spain and Egypt) includes many ideas in his great law code, the *Mishneh Torah*, about bringing a sense of refinement and spirituality to the eating process. One should not gorge oneself while eating, but rather stop eating before one is full. (This would go a long way toward reducing the epidemic of obesity in Western countries.) He also points out that a majority of our illnesses are caused/exacerbated by overeating.

In his *Guide to the Perplexed (More Nevukhim 3:33)* Maimonides writes:

> It is also the object of the perfect Law to make a human being reject, despise, and reduce his desires as much as is in his power. He should only give way to them when absolutely necessary. It is well known that it is intemperance in eating, drinking, and sexual intercourse that people mostly crave and indulge in; and these very things counteract the ulterior perfection of humans, impede at the same time the development of one's first perfection, and generally disturb the social order of the country and the economy of the family. For by following entirely the guidance of lust, in the manner of fools, one loses intellectual energy, injures the body, and perishes before the natural time.... The cause of all this is the circumstance that the ignorant considers physical enjoyment as an object to be sought for its own sake. God in great wisdom has therefore given us such commandments as would counteract that object, and prevent us altogether from directing our attention to it, and has debarred us from everything that leads only to excessive desire and lust.

In today's world we eat too much food, unhealthy foods, and food that is prepared in unhealthy ways. These unfortunate patterns have shortened lives, spread disease, and helped foster an unhappy, sick population. For this reason, many of us are making gargantuan efforts to change the way we deal with food – from extracting it from the earth to our putting it in our mouths.

Sustainable eating includes consuming foods that keep us healthy to live a long, productive and creative life. In the next chapters you will find some simple actions we can take.

Number Twenty-Four

Eat Organic

The original intention of adding pesticides, herbicides and fertilizers to our growing food was to help reduce competition for nutrients from weedy species of plants, reduce the number of organisms that eat the seeds, leaves and roots of plants we value for food, and to provide a "nutrient boost" to our crops so they grow faster and bigger. We now know that pesticides cause all kinds of problems to our bodies and to the environment. They are also responsible for an enormous amount of greenhouse gas emissions.

> **Recommended Actions:**
> - Purchase organic food where available.
> - The most important 12 products to buy organic are: Apples, bell peppers, celery, cherries, imported grapes, nectarines, peaches, pears, potatoes, red raspberries, spinach, and strawberries.

These chemicals may be toxic to the nervous system or the endocrine (hormone) system. In addition, many of these chemicals, along with fertilizers, run off into our lakes, streams and groundwater, which are all sources for drinking and irrigation.[115] Pesticides and herbicides also often kill what scientists call "non-target species" that are often not even threats to the crops of interest.

In addition to helping to heal the planet, eating organic will reduce our own chemical exposure. Healing the planet begins with healing our own bodies.

The Centers for Disease Control and Prevention (CDC) has detected pesticides in blood and urine samples from nearly 96 percent of more than 5,000 Americans tested in the agency's national biomonitoring program.[116] More than fifty percent of Americans tested carry in their bodies 10 or more fruit and vegetable pesticides and pesticide metabolites on any given day.[117]

According to the US Department of Agriculture, pesticides are detected in 7 of every 10 fruit and vegetable samples tested.[118]

Based on an analysis of more than 100,000 U.S. government pesticide test results, *Consumer Reports* recommends that the following 12 fruits and vegetables be purchased as organic produce, especially if they will be fed to children: Apples, bell peppers, celery, cherries, imported grapes, nectarines, peaches, pears, potatoes, red raspberries, spinach, and strawberries. *Consumer Reports* writes:

> *The U.S. Department of Agriculture's own lab testing reveals that even after washing, some fruits and vegetables consistently carry much higher levels of pesticide residue than others. Based on an analysis of more than 100,000 U.S. government pesticide test results, researchers at the Environmental Working Group (EWG), a research and advocacy organization based in Washington, D.C., have developed the "dirty dozen" fruits and vegetables {cited above}, that they say you should always buy organic if possible because their conventionally grown counterparts tend to be laden with pesticides.*[119]

(For more information on pesticide levels for other types of produce, go to www.foodnews.org.)

Foods labeled "USDA Certified Organic" are certified, by the United States Department of Agriculture,[120] that the ingredients have not been genetically modified, and were grown on land free of chemicals for at least three years. Certified organic livestock has not been injected with antibiotics or hormones.

Today many supermarkets sell a range of organic products, so it's become easier to purchase healthy foods there. You can also shop in local health food stores where organic products are plentiful.

Put it into Action!

Where can I find organic food?

Which foods in my diet are most likely to have high levels of pesticides?

What types of products will I start buying organic?

Number Twenty-Five

Buy Local

"Better a neighbor who is near than a brother who is far away." Proverbs 27:10

Recommended Actions:
- Find a local CSA and/or farmers' market.
- Look for ways to introduce local food to your diet.

Another way to make your food habits more sustainable is to buy "local food," foods that are produced as close to home as possible. When you buy local food, you support nearby farmers and the local economy, and ensure that you food doesn't have a global trip to your table, which saves energy. Because it didn't take days to get to you, local food is often fresher and tastes better. People who make an effort to buy local food are sometimes called "locavores."

One way to buy local is to participate in "Community Supported Agriculture" (CSA), which is basically purchasing a share in a local farm. Participants purchase a share of a local farmer's produce at the beginning of a growing season, and in exchange receive a box of farm-fresh seasonal produce each week of the growing season. There are a range of different types of CSAs – sometimes the food is organic, sometimes not, and the price, produce, and growing season vary widely. You should look for a CSA that works for you.

One thing that is always true about a CSA is that it's hard to predict how much food you will receive during a given season. That's because the amount of food the farmer produces is always dependent on the weather! By paying a fixed rate, you share in the risk with the farmer.

You can find the closest CSA near you by going to www.foodroutes.org. The Jewish-environmental organization Hazon also organizes CSAs in the Jewish community.[121]

There are now over 1000 CSA farms across the US and Canada, and the numbers are increasing at a significant pace. Cathy Roth, on the web site http://www.umassvegetable.org/food_farming_systems/csa/ adds these reasons for the importance of CSA's:

Why Is Community Supported Agriculture Important?
- CSA encourage direct communication and cooperation among farmers and consumers.
- CSA provide farmers and growers with a fair return on their labor.
- CSA keep food dollars in the local community and contribute to the development and maintenance of regional food systems.
- With a "guaranteed market" for their produce, farmers can invest their time in doing the best job they can producing food rather than marketing their products.
- CSA support the biodiversity of a given farm and the diversity of agriculture.
- CSA create a sense of social responsibility and stewardship of local land.
- CSA put "the farmers face on food" and increase understanding of how, where, and by whom our food is grown.

In addition to all these benefits, by buying into a CSA you are also developing a relationship with your local ecosystem. The kitchen table will become a more direct reflection of local climates and species. Many CSA farms also welcome their purchasers to visit and work an afternoon on the farm, a great experience for kids and adults alike.

Another option for buying local food is to seek out and buy products at local farmers' markets. Farmers bring the produce of their farm to a local, temporary market, usually set up outdoors in a lot or on a closed street, at a specific time each week during the growing season. At a farmers' market, you can meet the local farmers and hear from them about the produce they raise.

To find a farmers' market or CSA near you, search with your zip code at http://www.localharvest.org/ or www.eatwellguide.org.

Put it into Action!

Farmers' Markets Near Me (Days, Times, Location)

CSA Opportunities to Explore

Experiments with Local Food Purchases

SECTION H: SUSTAINABILITY AND JEWISH EATING 141

Number Twenty-Six

Eat Less (or No) Meat

And God said: "Behold, I have given you every herb yielding seed which is upon the face of all the earth, and every tree that has seed-yielding fruit -- to you it shall be for food." (Gen.1:29)

(See Rabbi Eric Yoffie's iniative, delivered at a meeting of Reform Jewish leaders, http://urj.org/life/food)

Recommended Actions:

- Access how often you eat red meat, chicken and fish.
- Look for ways to reduce animal proteins in your diet.
- Look for more sustainable meat choices.
- Consider becoming a vegetarian.

According to Jewish tradition, the first humans were vegetarians.[122] Nahmanides (13th century, Spain) wrote that the reason for this initial dietary law was:

Living creatures possess a moving soul and a certain spiritual superiority which in this respect make them similar to those who possess intellect (people) and they have the power of affecting their welfare and their food and they flee from pain and death.[123]

In the times of Noah, permission was given to eat meat.[124] And eat meat, we do.

Since 1960, global meat production has increased more than three times. Both population increases and greater affluence throughout the world are driving this trend. Global production and consumption of meat is expected to rise from the 233 million metric tons produced in 2000 to 300 million in 2020. Much of this increase is associated with a rapid rise in poultry production: 9 million metric tons in 1960, to 15 in 1970, 26 in 1980, 41 in

1990 and 68 million metric tons in 2000, thereby overtaking the production of beef (60 million metric tons in 2000). [125]

One reason for the increase in meat consumption is the rise of fast-food restaurants as an American dietary staple. As Eric Schlosser noted in his best-selling book *Fast Food Nation*, "Americans now spend more money on fast food—$110 billion a year—than they do on higher education. They spend more on fast food than on movies, books, magazines, newspapers, videos and recorded music—combined."[126]

While scientists have debated the pros and cons of a meat- vs. vegetarian-based diet,[127][128] the primary issue from an environmental point of view is the method of modern techniques for meat production. [129],[130]

Large-scale industrial farming operations are causes for multiple illnesses and disease.[131] The great majority of animals eaten in the United States are raised in Concentrated Animal Feeding Operations (CAFOs), agricultural operations where animals are kept and raised in confined situations, and which have waste materials that come into contact with the water supply. CAFOs congregate animals, feed, manure and urine, dead animals, and production operations on a small land area. Animals are confined for at least 45 days in a 12-month period, and there is no grass or other vegetation in the confinement area during the normal growing season.[132]

CAFOs cause significant environmental impacts to groundwater, surface water, air quality, and emit greenhouse gases that contribute to climate change. CAFOs and their waste can be breeding grounds for insect vectors, and pathogens found in animal manure can cause significant health effects.[133] Full-scale industrialized factory farming has given rise to recent, highly publicized epidemics of meat-borne illnesses.[134] CAFOs are sources of disease that affects humans, e.g., the H1N1 virus.[135]

The US Department of Agriculture estimates that large animal farms produce 575 billion tons of manure every year.[136] This waste is not confined to animal feeding operations. The 1995 New River hog waste spill in North Carolina poured 25 million gallons of excrement and urine into the water, killing an estimated 5000 fish[137] and closing 364,000 acres of coastal shellfish beds.

Energy-intensive U.S. factory farms generated 1.4 billion tons of animal waste in 1996, which, the Environmental Protection Agency reports, pollutes American waterways more than all other industrial sources combined.[138] Meat production has also been linked to erosion and to the destruction of rainforests.[139]

In the United States, 11.2 million kg of antibiotics are used annually as growth promoters for livestock; only 1.4 million kg are used for humans for all pharmaceutical purposes. In other words, 87% of all antibiotic use is for animals, while only 13% is for human therapeutic and non-therapeutic use.[140] "For sheer over-prescription, no doctor can touch the American farmer," reported *Newsweek*.[141] Using these antibiotics on animals reduces their effectiveness for human beings.[142]

Beyond health, eating meat has important consequences for world hunger. The 4.8 pounds of grain fed to cattle to produce one pound of beef for human beings represents a waste of resources in a world in which many still suffer from profound hunger and malnutrition. For example, 50% of the corn we raise is not for direct human consumption but for cattle; add poultry, fish and pigs and that number jumps to 80%.[143] As *Diet for a Small Planet* author Frances Moore Lappé[145] writes, imagine sitting down to an eight-ounce steak. "Then imagine the room filled with 45 to 50 people with empty bowls in front of them. For the 'feed cost' of your steak, each of their bowls could be filled with a full cup of cooked cereal grains."[146]

Being confined in animal feeding operations is no treat for the animals either. Livestock that are raised in CAFOs are subjected to a range of chemical irritants that cause adverse health effects. Ammonia from urine, hydrogen sulfide gas from the decomposition of sulfur-containing proteins and exposure to toxic microbial by-products are a few of the chemicals to which the animals are exposed.[147] They are often confined in cages or crates, or confined in indoor sheds for their entire lives. They may experience discomfort and injuries caused by inappropriate flooring and housing, the restriction of normal exercise and natural animal behavior (such as foraging or exploratory behavior), lack of daylight and fresh air, and health problems caused by breeding and management for fast growth and high productivity.

It should be mentioned that there are some small companies that are intentional eco-minded kosher meat providers, such as Grow and Behold (www.GrowandBehold.com). Their web site states:

Grow and Behold Foods brings you delicious pastured meats raised on small family farms. Our meats are produced in limited quantities to ensure that we adhere to the strictest standards of kashrut, animal welfare, worker treatment, and sustainable agriculture. We do it right, so you can enjoy every bite.

Another similar company is Kol Foods (www.KolFoods.com) which informs their web site readers:

KOL Foods produces 100% grass-fed, sustainable kosher meat. We put kosher meat and ethics on the same plate so you can feel good about the meat you eat. Every day we work to create a new food system that supports sustainable animal production, treats farmers and workers fairly, and improves the health of families and communities. As a values-based business, we produce food that is in harmony with nature, neighbors and tradition - all the way from farm to fork.

Excessive meat eating is not good for us, it's not good for the animals, and it's not good for the planet! For this reason, many people have chosen to go vegetarian or only eat organic, free-range meat.

Recently a group of young idealistic Jews who are concerned about sustainability, about health and about the planet, started a web site: www.veguary.org. There is much interesting material on the site, especially, Ross Gitlin's article, "Why I Became a Vegetarian."

If that feels like too big a step for now, you can have a measurable impact by simply replacing one steak, plate of spaghetti and meatballs, beef lasagna or a trip to the local fast food joint with a nice vegetarian meal once a week.

In light of what we know in the 21st century about the harmful effects on the human body of eating meat, as well as effects on the environment, eating less or no meat is something everyone might consider.

An Animal-Eating Record for the Month of _____ :

(Put a check each time you eat one of the following. See if you can eat lower on the food chain as the month goes on.)

Put it into Action!

Week of:	Beef	Chicken	Fish	Eggs/Milk

Figure 10

Why I am a Vegetarian

Richard H. Schwartz

Until about 1977, I was a "meat and potatoes" person. My mother was sure to prepare my favorite dish, pot roast, whenever I came to visit with my wife and children. It was a family tradition that I would be served a turkey drumstick every thanksgiving. Yet, I have not only become a vegetarian, but I now devote a major part of my time to writing, speaking, and teaching about the benefits of vegetarianism. What caused this major change?

In 1975, I began teaching a course, "Mathematics and the Environment" at the College of Staten Island. The course uses basic mathematical concepts and problems to explore current critical issues, such as pollution, resource scarcities, hunger, energy, and the arms race. While reviewing material related to world hunger, I became aware of the tremendous waste of grain associated with the production of beef. (Over 70% of the grain produced in the United States is fed to animals destined for slaughter, while an estimated 20 million of the world's people die annually due to hunger and its effects.) In spite of my own eating habits, I often led class discussions on the possibility of reducing meat consumption as a way of helping hungry people. After several semesters of this, I took my own advice and gave up eating red meat, while continuing to eat chicken and fish.

I then began to read about the many health benefits of vegetarianism and about the horrible conditions for animals raise don factory farms. I was increasingly attracted to vegetarianism, and on January 1, 1978, I decided to join the International Jewish Vegetarian Society. I had two choices for membership: (1) practicing vegetarian (one who refrains from eating any flesh);

(2) non-vegetarian (one who is in sympathy with the movement, while not yet a vegetarian). I decided to become a full practicing vegetarian, and since then have avoided eating any meat, fowl, or fish.

Since that decision, I have learned much about vegetarianism's connections to health, nutrition, ecology, resource usage, hunger, and the treatment of animals. I also started investigating connections between vegetarianism and Judaism. I learned that the first Biblical dietary law (Genesis 1:29) was strictly vegetarian, and I became convinced that important Jewish mandates to preserve our health, be kind to animals, protect the environment, conserve resources, share with hungry people, and seek and pursue peace all pointed to vegetarianism as the best diet for Jews (and everyone else) today. To get this message to a wider audience I wrote a book, Judaism and Vegetarianism, which was published in 1982. (new expanded editions were published in 1988 and 2001.)

Increasingly, I have come to see vegetarianism as not only an important personal choice, but also as a societal imperative, an essential component in the solution of many national and global problems. The U. S. Surgeon General has indicated that 68% of diseases in the United States are related to poor diets, and this is a major factor behind soaring medical expenditures, a key reason for the tremendous debt the U. S. faces. Also, livestock agriculture is a major contributor to many current environmental and public health threats, such as climate change, the destruction of tropical rainforests and other habitats, soil erosion and depletion, water shortages, air and water pollution, and the proliferation of antibiotic-resistant, disease-causing bacteria.

I have recently been spending more and more time trying to make others aware of the importance of switching toward vegetarian diets, for them and for the world. I have: appeared on over 60 radio and cable television programs; had many letters and several op-ed articles in the Staten Island Advance and other publications; spoken frequently at the College of Staten Island and to community groups; given many talks and met with three chief rabbis and other religious and political leaders in Israel, while visiting my two daughters and their families. In 1987, the Jewish Vegetarians of North America (JVNA) selected me as "Jewish Vegetarian of the Year" and in 2005 I was inducted into the North American Vegetarian Society's Hall of Fame. Recently, I have been using email and the Internet to continue my efforts to help increase awareness of vegetarianism.

As part of my activism, I am President of Jewish Vegetarians of North America (and editor of their newsletter) and the Society of Ethical and Religious Vegetarians (SERV) and director of Veg Climate Alliance (VCA), an organization dedicated to increasing awareness of connections between animal-based diets and climate change. In 2000 I became a dietary vegan.

I have always felt good about my decision to become a vegetarian. Putting principles and values into practice is far more valuable and rewarding than hours of preaching. When people ask me why I gave up meat, I welcome the opportunity to explain the many benefits of vegetarianism.

While my family was initially skeptical about my change of diet, they have become increasingly understanding and supportive. My wife is vegetarian as are my younger daughter and several of my grandchildren. In 1993, my younger daughter was married in Jerusalem at a completely vegetarian wedding.

There is much that still needs to be done. My hope is to be able to keep learning, writing, and speaking about vegetarianism, to help bring closer that day when, in the words of the motto of the international Jewish Vegetarian Society, "no one shall hurt nor destroy in all of God's holy mountain." (Isaiah 11.9)

Richard H. Schwartz
Professor Emeritus, College of Staten Island
Author of *Judaism and Vegetarianism, Judaism and Global Survival* and *Mathematics and Global Survival* and over 140 articles at JewishVeg.com/schwartz
President. Jewish Vegetarians of North America (www.JewishVeg.com);
Director of Veg Climate Alliance
Associate producer of A SACRED DUTY
president@JewishVeg.com

Number Twenty-Seven

Sustainable Meat Options

Vegetarianism is a good choice for some, but it is clear that many people will continue to eat meat. And, in fact, despite the numerous sources shared in the previous section, today kosher meat eating is permitted by Jewish law.

If you just can't cut meat out of your diet, there is a growing market for sustainable kosher meat which might help you live your values while also satisfying your palette.

A recent blog posting at "The Jew and the Carrot"[148] commented on the "growing demand for meat that is local, humanely raised and ethically slaughtered," and provides a detailed list of resources to help you find a provider of sustainable kosher meat that will serve your needs. This listing is for companies which serve New York City, but many ship beyond that region and even nationwide (or course, removing the "local" aspect). The page also includes many resources related to Jewish thought and meat-eating (or not eating).

If you are shopping for a sustainable beef option, ask if animals are fed hormones, antibiotics, animal bi-products or arsenic (used to speed up the growth of conventional poultry). You can also ask if they live in "combined animal feeding operations" (CAFOS), warehouses where animals are rarely exposed to sunlight and fresh air, or if they have the opportunity to get outside. In addition to the suffering to the animals caused by overcrowding, CAFOs also generate enormous amounts of waste that can pollute local water resources.

Many of these "sustainable kosher meat" ventures feed their animals grass instead of the usual corn feed. Grass-fed cows eat grass their entire lives, as cows have evolved to do. These cows are also often raised as part of a farm ecosystem, in which cows are rotated through various pastures and their wastes are used as fertilizer and even food (think: flies which feed chickens) for other farm products.[149]

Some "sustainable kosher meat" is certified organic. According to the USDA National Organic Program, meat, poultry, eggs, and dairy products bearing the USDA organic label come from animals that are given no antibiotics or growth hormones, and that are fed organic feed. Livestock must also have access to the outdoors.[150] Whether the livestock actually *graze* outdoors may be another question.[151]

While the USDA organic label certifies compliance with national organic standards, it is important to remember that some smaller or local farmers may find these standards and the certifications to be unnecessarily expensive or difficult to demonstrate, even if they are raising their animals in a healthy and sustainable way.

When considering your meat choices, it's important to identify which factors are most important to you. For example, are you most concerned with:
- Health of the animals?
- Taste of the meat?
- Treatment of animals?
- Environmental impact?
- Health impact for yourself?
- National certification?
- Shared values?
- Kosher status?

As you consider your options, get to know the distributors or even the farmers that provide this meat. It's a small industry, and individuals will be interested and responsive to your concerns. Find out which standards they are keeping to, and why they've made their choices. You can then purchase meat from someone who matches your values.

Put it into Action!

Kosher Sustainable Meat

What matters most to me?

Companies to consider:

Purchases made:

Number Twenty-Eight

When You Eat Fish

One of the ten regulations established by Joshua for settling in the Land of Israel was that non-residents of Tiberias were allowed to fish in the Kinneret with fishing rods only and not nets. Talmud Bava Batra 81a.

Oceans cover nearly 70% of the earth's surface, and are the basis of the global food chain, providing food for birds, larger fish, and other mammals that keep our ecosystems thriving. Of course, the oceans also provide a significant source of food for human beings.

Recommended Actions:

- Access the sustainability and pollution levels of the fish you normally eat.
- Make a commitment to make healthier and more sustainable fish choices.

The intricate web of life in our oceans requires protection from numerous threats and pressures that are causing significant ecological problems. We need to eat fish that are sustainably harvested, and also avoid fish which may carry high concentrations of mercury and other toxins.

Over-fishing is a major threat to marine life.[153] Our technology and ability to catch fish from the ocean has rapidly improved, resulting in greater catch efficiency and ultimately more fish being taken from the sea. In fact, there are too many fish being taken from the sea, leading to a rapid decline, and in many cases, depletion of fish stocks. Worryingly, a 2006 study in *Science* predicted the global collapse of all fisheries by 2048.[154]

Another concern is that many species that are not targeted by fishermen still end up on their hooks or in their nets unintentionally.[155] This is known as bycatch. These include sea birds, sea turtles, whales, dolphins, sharks and a myriad of fish species not targeted by fishermen. Bycatch is of great concern since it leads to the major disruption of the food chain and the ocean's ability to thrive and provide a food source for all species (including humans).

Although the oceans may seem limitless and pure, in fact, they are being polluted.[156] The seafloor is littered with debris ranging from household trash to drums of discarded nuclear waste. Discharge pipes from factories, runoff from agricultural and farm lands, residential lawn fertilizers, and of course sewage outfalls, accumulate in the ocean. These toxins accumulate in the fish that we eat.

Another concern is the mercury emitted into the air that concentrates in marine organisms.[157] This mercury comes from coal-fired power plants which emit tons of methyl mercury into the air – a byproduct of the coal burning process. This mercury is eaten by tiny organisms who are eaten by small fish. These small fish are then eaten by larger fish that are then eaten by even larger fish. Ultimately, the fish at the top of the food chain like tuna, swordfish and other species can have high concentrations of mercury in their tissue. The problem is so pervasive that the EPA has issued mercury warnings for pregnant women and children who eat fish and other seafood.

We should take care that whenever we are eating fish, we make sure that they are not polluted with mercury or other toxicants – and also to make sure they are not being fished to extinction.

The Monterey Bay Aquarium Seafood Watch creates science-based recommendations that help consumers and businesses make ocean-friendly seafood choices. They offer a "Super Green List," which identifies fish that are good for you and good for the oceans.[158] They also offer handy regional and national printouts that you can place in your wallet and bring with you to the grocery store, and even mobile applications for your smartphone.

Put it into Action!

Choosing Sustainable Fish			
Fish I eat at least 3x per year	Monterey Bay Ranking	Mercury Concern?	Replacement options

Section I: Be a Green Jew

Number Twenty-Nine

Make Your Jewish Life Cycle Celebrations Green

Some of the most important occasions in Jewish Life are the life cycle observances, for birth, bat/bar mitzvah, weddings, and funerals. Each year, the community spends millions of dollars on invitations, food, music, and venues for these important events. Unfortunately, the events themselves do not always match with our Jewish values. They often generate enormous waste and, taken together, they represent a significant impact on the environment. At the time we're planning a "simcha" we're in a unique position to spend our resources carefully and to share our values with the community as a whole.

> **Recommended Actions:**
> - Use the tips in this section to integrate environmental consciousness into your next simcha.

You can make choices that will be better for your family, your community and the world — and by simplifying, most of these choices can even save you money.

Jews United for Justice offers a great guide to "Green and Just Celebrations" which can help you think through all the details. The book is available free for download or for inexpensive order on their website at http://www.jufj.org/green_just_celebrations.

Bar and Bat Mitzvahs are a special opportunity to help young men and women connect to their Jewish values and understand how Jewish tradition can inform the most important challenges of today. As you prepare for your child's Bar or Bat Mitzvah, you have a unique opportunity to help your child grow and learn as a Jewish adult. By making the celebration reflect your values, you can make a long-lasting impression.

For example, an article in the Boulder Jewish News from 2009[159] provides great tips for a green Bar or Bat Mitzvah, such as:
- Less wasteful invitations
- Eco-friendly Kippot
- An environmental "Mitzvah project"
- Tutor nearby
- Greener flower arrangements
- Zero Waste
- Useful decorations

For the full article, see Figure 11.

Weddings are also a great opportunity for a young couple to demonstrate their values. Many young couples go a bit overboard with the planning of one fantastic day, and then realize a year later that they could have used the money for a much better purpose. Eco-friendly weddings can be simpler, less expensive, and more distinctive, leaving both guests and hosts with a sense of pride and satisfaction.

In "Enviro-Chuppah," an article on the Canfei Nesharim website, writer Yaakov Lehman notes that environmentally friendly weddings are becoming an increasingly popular option across the globe. He recommends the following websites to help you:
- Greenweddings.net: offers a variety of consultation and contracting services in the realms of green transportation, decorations, gowns, photographers, invitations, wedding rings, and a host of others.
- Portovert, "the modern bride's guide to a green wedding"
- Organicbouquet.com, a source for environmentally-friendly flowers.

The article also recommends the following simple actions to make your wedding more green:

1. Print wedding invitations on 100% PCW chlorine free recycled paper.
2. Use RSVP postcards to save paper on envelopes.
3. Purchase sustainable wedding garments, or a dress/suit that is practical to wear on other occasions.

4. Suggest donations to your favorite environmental organization for gifts.
5. Host your ceremony and reception at the same location.
6. Purchase local and organic produce for your reception.
7. Use reusable dishware, or purchase biodegradable/compostable dishes and cutlery made from cornstarch.
8. Educate your guests about your eco-friendly choices, though your wedding booklet, bentchers (booklets containing Grace After Meals), and other gifts.

A recent article from the Rodale.com Daily Newsletter offers some valuable tips for "green weddings", with some basic rules:

- Give up on perfect.
- Use recycled and reusable materials.
- Pick your own flowers.
- Use the ceremony as an opportunity to educate.
- Opt out of certain wasteful traditions.
- Set a sustainable table.
- Forgo favors!

For the full article, see Figure 12.

Jewish National Fund encourages families to use JNF Tree Certificates as invitations and favors, so that not only are you inviting your friends and family to celebrate your joyous occasion but you are also planting trees in Israel to improve Israel's environment and stop desertification. Contact http://www.jnf.org/get-involved/celebrate/

When the time comes for your Simcha (brit milah, bat mitzvah, graduation, wedding, birthday, anniversary) be creative and find ways to share your commitment to a more sustainable world.

Put it into Action!

Ideas to Green an Upcoming Celebration or lifecycle event:

Figure 11

7 Ways to Green a Bar/Bat Mitzvah

By Debbie Garelick

Boulder Jewish News - December 1, 2009

© Boulder Jewish News, Inc. 2009-11. Used with permission

We do not inherit the earth from our ancestors; we borrow it from our children.

~Native American Proverb

The B'nai Mitzvah year is one of learning, planning, shlepping, and growing. We want our kids to become Jewish adults and understand how to continue to grow and learn as a Jewish adult throughout their lives. We teach them Torah, prayers, customs and HOPEFULLY we teach them to give back through tzedakah and to help repair what is wrong with the world through Tikkun Olam.

We can start repairing the world during this transition from Jewish childhood to Jewish adulthood. Here are 7 **great tips** for greening the Bar Mitzvah year, the service and celebrations:

1. **Start with the invitations.** You have a date for the event. Plan ahead and use recycled paper for your invitations. You can even get paper with seeds in it so that when the recipients are done with it, they can plant it in their garden and grow flowers or veggies. Less is more. Try not having as many envelope [sic]: many people now make the RSVP card a postcard (thus no envelope) or even better, have guests RSVP by email. Remember to have a phone number to RSVP as well for those few people who do not have an email address. You can make your own

invites for less money as well. If you print it on your printer, try downloading free eco-friendly fonts (http://www.ecofont.com/en/products/green/font/download-the-ink-saving-font.html) that use less ink. Of course you can be completely green by making an online-only invitation. Choose your green limit and push the envelope!

2. **Tutor nearby.** If a tutor is part of the plan, choosing the tutor for your kids is a very important decision. Are the personalities right, the level of learning a good match? You can also look to see if someone is close by or near a bus route from your child's school or your home. Maybe ask your kids to bike to their lesson on warmer days.

3. **Ordering kippot can be green.** Eco Suede [http://zaramart.com/p1150.htm] makes a kippah from recycled cardboard. Fair trade kippot from Guatemala are also a wonderful choice.

http://shop.mayanhands.org/Fair-Trade-Crocheted-YarmulkeKippah--NEW_p_276.html

4. **The Mitzvah project is a natural fit.** Doing a "Mitzvah project" has become a wonderful part of the B'nai Mitzvah process. There are many wonderful projects that kids can do to help the world. If your child has a leaning towards helping the environment, they may be inspired by a child in our community who, this past summer, helped fund and organize solar panels for the Boulder Humane Society. Per Dustin Michels,

age 12 at the time, "I helped the Humane Society to go green through solar because I hoped it would make an impact on the environment, the community, and their electrical bill." WOW!

5. **Think trees instead of flowers.** Instead of ordering flowers for the Service at the synagogue, you could buy a plant or small tree to plant after the festivities. If it is the middle of winter, you can make a sign that says you will plant a tree at the synagogue in the spring. This can be even less expensive than a fancy arrangement and "green" the synagogue for many many years. If it becomes a tradition for a synagogue, a small B'nai Mitzvah garden can be planted and new Jewish adults can be represented by planting their own trees, contributing to the synagogue and to the planet!

6. **Make your event Zero Waste.** Ok, the day has arrived after all the studying and planning. The Torah, Haftarah, and Drash (speech) are perfect and it is time to celebrate. You can go to <u>Ellie's Eco Home Store</u> next to <u>Sunflower Market</u> and purchase all the plates, flatware and cups in compostable form. Don't forget the compostable bags for the trash. Made from either corn or potato, these products can go into the compost instead of the landfill. Landfills produce methane gas which is 7 times more harmful to the climate than carbon. Try to buy bulk food products or products with less packaging. And although, I admit, I am a meat eater, so I try to limit my meat and fish intake. Plus, with Kashrut laws vegetarian is SO MUCH EASIER for the B'nai Mitzvah celebrations. I'll post some great vegetarian Indian Kiddush recipes in an upcoming article. Also, think about using locally grown food.

7. **Decorations with more use.** When thinking about decorations, think about making something that can be donated to the needy afterwards. Recently I went to a Bat Mitzvah where the center pieces were new men's socks and gum balls. The gum was eaten during the celebration and the socks were donated to the homeless shelter where they are in great need. Instead of decorations that will be thrown away, they were used for a worthy cause. It can also be tied into the Mitzvah project your child has chosen.

Make this special time in your families lives a memorable and green one. We are all on our own special Jewish and green path. Choose the path that is right for you and your child. And in the process make a difference and repair the world. Please share your ideas, and teens – we'd love to hear about your mitzvah projects!

Debbie Garelick

Debbie is a Realtor, business woman, mom of two teens and wife of Robert (almost 23 years). She is a strong advocate for local environmental issues in the Jewish community. Also, she is an advocate for juvenile diabetes. Debbie has lived in Boulder since 1974 except for 6 years when she lived in Israel residing on Kibbutz Maagan Michael.

Figure 12

Rules for Greening Your Wedding

Plan a green wedding: Keep hundreds of pounds of garbage out of the landfill, reduce air pollution, and save serious cash in the process.

By Leah Zerbe

The article below is from the web site www.Rodale.com (Rodale.com Daily Newsletter, May 27, 2010)

RODALE NEWS, EMMAUS, PA—Wedding season is in the air, but after all the dancing, champagne toasting, and celebrating, what's left behind is often an ugly mess—up to half a ton of garbage, food trucked in from hundreds of miles away that winds up being wasted, and junky favors that wind up in the landfill. (Seriously, does anyone really use those gaudy shot glasses with the bride and groom's wedding date printed on it? Or the toxic candles or soaps, or personalized plastic trinkets made in China?)

Green weddings can save tons of carbon pollution and keep tons of garbage out of landfills—a major positive in itself—but you can also enjoy some immediate benefits of greening your wedding. At the top of that list? Saving mega money. "Green weddings have absolutely moved mainstream, which is really exciting," says Kate Harrison, author of *The Green Bride Guide: How to Create an Earth-Friendly Wedding on Any Budget* (Sourcebooks Casablanca, 2008). "Less impact on the Earth is a major motivator, but another is the desire to save money."

THE DETAILS: A recent David's Bridal survey found that 46 percent of couples are thinking about incorporating ecofriendly components into their wedding festivities. That's great news, because the average wedding creates 400 to 600 pounds of garbage and 63 tons of carbon pollution. One year of American weddings creates about the same CO_2 pollution as putting 8 million cars on the road.

WHAT IT MEANS: It's easy to associate a green wedding with more expense, but if you play your cards right, you can actually save money and slash your celebration's carbon footprint, too. Author <u>Maya Rodale</u> director of communications and outreach at the <u>Rodale Institute</u>, an organic research farm in eastern Pennsylvania, is in the midst of planning her own green wedding. Her upcoming historical romance novel, *A Groom of One's Own* (Avon, June 2010), even features components of a green wedding. "There was no question that I would have an organic wedding—given my family's history and the values I live by, it was inevitable. And I'm excited by the challenge," says Rodale. "Plus, I am excited for the opportunity to show friends and family how gorgeous, delicious, and lovely organic is."

Here are some rules to live by when greening your wedding:

- **Rule No. 1: Give up on perfect.** When greening a wedding, always remember it's not all or nothing, and every eco-upgrade you make is helping the human race and the planet. "Make the better, greener choices where you can, and don't stress what you can't," suggests Rodale. "My menu will be totally organic because I know a great caterer who does that. My dress is not at all organic because I didn't love any of the eco-gowns I found."

- **Rule No. 2: Use recycled and reusable materials.** Reusing something is the ultimate in eco-awareness, since it keeps items out of the trash and spares the energy used to make, package, and transport something new. But that doesn't mean you have to shop at a trendy eco-boutique. Here are a few ideas:

1. Old, unique bottles found at thrift stores or antique shops, or even the junkyard make eclectic centerpieces to hold local wildflowers.

2. Reusable tablecloths, aisle runners, and napkins from similar sources can add some vintage appeal. (You can even make them out of old tablecloths found in thrift stores. <u>It's easy!</u>)

3. Look for a preowned gown or wedding shoes. Since they've only been worn once, these are practically new. Harrison was able to buy once-worn Vera Wang shoes for a fraction of the original price on eBay. Then, she resold them for $50. "You can do the same thing with a wedding gown, veil, or other accessory," she says. "When you're finished, you can donate, sell, or recycle it." And of course there's the tradition of wearing the gown that belonged to your mother or another relative. Even if you have it altered or updated, it counts as reuse.

4. Rodale is renting nice wooden farm tables, which won't need to be covered with linens. "We'll save on tons of toxic dry cleaning there. It's a little thing, but it still counts," she says.

- **Rule No. 3: Pick your own flowers.** Harrison, a Yale School of Forestry graduate with a J.D. in environmental law, was married in the fall in the Hudson Valley in NY. "I wanted my wedding to feel like the season it was held in," she says. So instead of opting for expensive, exotic flowers shipped in from another country, she saved lots of money by selecting hydrangeas and dahlias from a local organic farm. She also collected ferns and pine cones locally and handed them over to the floral designer to create beautiful, local arrangements. Wedding flowers typically cost about $2,000, but Harrison only wound up spending about $200.

- **Rule No. 4: Use the ceremony as an opportunity to educate.** Nothing could be more off-putting than a preachy Bridezilla. But there are tasteful ways to educate your guests on green issues without being overbearing. You may choose to use local native plants as centerpieces, with a little note explaining how these flowers benefit birds, bees, and butterflies in your area.

If you don't want to publicize your green-ness at your wedding, you could post information under a tab on your wedding website. "That way, it's there for anyone who wants to learn more," explains Harrison, who also launched a website full of green wedding ideas, (http://www.greenbrideguide.com/) including a green registry and green honeymoon ideas. You could also add an eco-awareness note on your seating cards or in your program. It could even be as simple as using a recycled symbol on your recycled paper invites or programs.

- **Rule No. 5: Opt out of certain wasteful traditions.** Consider making your own invitations and printing them on recycled paper. And please, forgo all the extra envelopes and tissue paper. "I have read the reason was to blot the ink and keep everything clean in the days when they were handwritten," says Harrison. "Obviously, this is not needed anymore, so it is just a relic."

Rodale is avoiding this type of waste by using a self-mailer RSVP card instead of cards*and* envelopes, and they are printed on recycled paper. "Rehearsal dinner invites will be via paperless post, a beautiful online invitation option," she adds.

- **Rule No. 6: Set a sustainable table.** Factory-farmed food is cheaper because it is subsidized by the government, but the animals live in horrid conditions, are fed antibiotics that lead to superbug outbreaks, and pollute drinking-water sources and our air. Chemicals used on crops degrade the health of the soil and can end up on, and in, the food we eat. So in the end, cheap chemical food costs a lot more.

For Rodale, food definitely took precedence over other wedding greening categories. "Our menu will be all-organic, all-local. It's more expensive, so we're cutting costs in other ways to compensate," she explains. "Organic food is one way to make a major, beneficial impact on the environment and local economies. Plus, everyone can appreciate a darn good meal, and what better way to showcase organic?"

If your wedding is slated for next spring, summer, or beyond, consider growing some of your own wedding food. You can grow more than 100 pounds of produce in an average-size backyard garden. For more tips on growing, visit OrganicGardening.com.

"We're growing our own flowers and food to supplement our menu and decorations. We have a farm and even we can't grow it all, so I do not recommend anyone else try to do so," Rodale says. "But growing a bit can help with costs. We're also having our wedding on the farm, so that the décor will definitely be all-natural, all-organic."

Also, work with your caterer. He or she may allow you to source in-season, organic produce through a local farmer. Buying direct could cut costs, and if you buy in bulk, it puts you in a better position to negotiate for a lower price. The farmer could also cut you a deal if you mention the farm in your wedding program. Visit LocalHarvest.org to find a grower in your area.

- **Rule No. 7: Forgo favors!** Another great way to green your wedding and save money? Nix table favors. Saving this extra cash could allow you to splurge on a great local band or more sustainable food. You could also make a donation to the nonprofit of your choice in lieu of a favor. "For favors, we're giving folks a darn good dinner, lots of dancing and, hopefully, a night to remember," says Rodale. "So we're not providing landfill-clogging goodies. Sorry!"

Number Thirty

Green Your Jewish Holidays

The Jewish holiday cycle includes numerous opportunities to learn about Jewish sources that connect us to the land. For example, the holidays of Pesah, Shavuot and Sukkot reflect an ancient harvest cycle, in addition to their historical meanings.

Rabbi Yitzchak Breitowitz[160] explains that while Pesah celebrates the exodus from Egypt, Shavuot celebrates the giving of the Torah, and Sukkot commemorates our wandering in the desert, Pesah is also the beginning of the spring when the barley was harvested. Shavuot was also the time of the harvesting of the wheat, and Sukkot was the ingathering of the produce and fruit. He continues explaining that in a Jewish leap year, we add an additional Jewish month of Adar so that Pesah will always occur in the spring. He writes:

> **Recommended Actions:**
> - Learn about the environmental linkages to Jewish holidays.
> - Integrate environmental lessons into your observance of Jewish holidays.
> - Reduce the environmental impact of your observance of Jewish holidays.
> - See the national Jewish movements web sites for more information on their environmental efforts.

> *Why is it so important that Pesah be in the spring? The gemara says there's an intimate bridge between the physical structure of the universe and the spiritual universe. What happens on this earth models the "spiritual vibe" that God is putting into the earth at that time. For example, take freedom. That will be manifested by the fact that the earth itself becomes liberated after a long cold winter, after it was dormant, to begin with productivity and growth. Pesah occurs in spring because that's the time of freedom.*

Shavuous, the time of harvest, is also the time of the matan torah, giving of the Torah, when we can harvest the knowledge that Hashem has made available to us. Sukkos is the time of ingathering; so, too, after Rosh Hashanah and Yom Kippur, we are able to internalize that knowledge into our emotions and behavior, ingathering into ourselves.161

Many Jewish holidays offer an opportunity to acknowledge our responsibility to care for the world, and in celebrating them, we often have an opportunity to reduce our environmental impact. For a full set of Torah teachings about the environment for each Jewish holiday, visit http://www.canfeinesharim.org/learning/holidays.php.

What follows are three creative examples of how people who are environmentally conscious can bring innovative, yet in a way also traditional, ideas to bear when celebrating Jewish festivals: Purim, Pesah and Sukkot.

The Israeli web site www.svivaisrael.org ("The Israel Environment") sent out the following list of ten suggestions for making Purim Eco-friendly. This list can be used as an example of how any and every Jewish holiday can be made environmentally sustainable.

10 Tips for an Eco-Friendly Purim

Purim celebrates unity and community. We give one another gifts of food and distribute money to the poor to remember how we were rescued from Haman's plot to destroy the Jewish people by pulling together in three days of communal fasting, prayer and self-reflection.

Sadly, in today's consumer-oriented society, our community spirit is being smothered in layers of cellophane, excessive packaging, a surplus of junk food and expensive costumes. This needless glitter is not only damaging to our pocket and our health, but to our environment as well. Sviva Israel offers the following suggestions for creating a more environmentally friendly Purim:

1. Trash the Baskets – What can you do with so many straw baskets and gift bags? Package your Mishloah Manot in useful, reusable containers such as storage containers, glasses, mugs and pasta drainers for year-round usability.

2. Wrap it Up – For the more creative, wrap your food items up in a pretty hand-towel, apron, cloth table napkins, oven mitts or other useful fabric item.

3. Sustainable Stuffing – Instead of padding out your package with shredded cellophane or colored paper, use banana chips, sunflower seeds or popcorn (only for recipients over 3-years –old).

4. Bag It – Follow the fashion trend and give your gifts in eco-friendly cloth bags that your friends can reuse for shopping.

5. Naturally Sweet – Replace the candy and chocolates with fresh and dried fruit or fruit leathers, unsweetened fruit juices and other healthy products.

6. Purim Swap Shop – Your son doesn't want to wear last year's cowboy outfit? Many costumes are perennial favorites. Create a neighborhood swap shop with everyone's unwanted, worn-once Purim costumes.

7. Raid Mom's/Dad's Closet – Introduce your kids to the old Purim tradition of creating their own costumes from your (old) clothing, hats, shoes and jewelry. Encourage their imagination to run wild!

8. Recycling can be cool – Making a costume from cardboard boxes, kitchen roll tubes etc. needn't be old-fashioned. Your child could become an ipod, cellphone or XBox!

9. Join a Purim Co-op – Give Mishloah Manot as a community. Compile a list of all the members in the community (neighborhood, synagogue, seniors group etc.). People can check off the names of those they would like to send a gift to, contributing a set amount for each name. Volunteers prepare and deliver ONE nice-sized food gift to each person, with

a note listing all of their friends who thought of them. The beauty of this idea is that is saves the time and excess food and packaging of multiple gift giving, creates a strong sense of community fellowship and any profits can be given to charity.

10. Share the Spoils – Purim is over and you find yourself overloaded with unwanted food gifts? Bring (unopened) food items to a local charitable organization to distribute to needy families.

The second idea comes from a small group of extremely inspired, inventive and resourceful students at The Jewish Theological Seminary in New York, who created an internet brochure on their school's web site, with suggestions about celebrating Pesah. (These "Eco-reps" from JTS are Melanie Schwartz, Renna Khuner-Haber, Adi Segal, Mara Berde and Becca Farber).

Recycling Pesah Customs

<u>Eco-Reps</u>, students at The Jewish Theological Seminary

Every family has its own customs for the Pesah Seder. Some of these customs can link Passover's idea of human liberation and environmentalism's idea of reconnecting humanity with the Earth. For instance, many Jews used to (and many still do) save their *lulav* from Sukkot to use to burn the hametz (leaven) before Pesah. In this way, we reuse one ritual object which was used for a mitzvah (commandment) to do another commandment. This custom, which originated in the 12th century, applies the concept of bal tashkhit (do not wantonly waste any part of Creation) – a key principle of Jewish environmental thinking. The custom also spiritually connects two festivals of Creation: one in the fall (Sukkot) and one in the spring (Pesah). It reminds us how everything is recycled through the rhythms of life, death, and rebirth.

Another unusual custom which dates back to the 13th century is to put some earth, clay or ground stone into the *haroset*, the mixture of nuts and fruit which is meant to remind us of the work of the Israelite slaves in making mortar for the bricks. During the Civil War, a group of Jewish Union soldiers made a Seder and put a brick on the Seder plate because they did not have the ingredients for *haroset*. While you may not want to put clay, earth, or stone into your *haroset*, you may want to put a stone on your Seder plate to symbolize slavery and the misuse of the earth in oppressing people. It can remind us that everything we own and eat comes from the Earth. Just don't break your teeth!

Modern Day Plagues

1. Pollution everywhere, and not a drop to drink (blood in rivers)
2. Ecosystems going "out of whack" (affecting frogs, and other living things)
3. Huge dislocations in daily life (think lice, but much worse)
4. Species extinction (beasts no more)
5. Loss of reliable food supply (cattle plague)
6. Spread of tropical diseases (boils and other globalizing ailments)
7. Changes in weather patterns (such as hail)
8. Crop failure (locusts and other agricultural nightmares)
9. Rising sea levels pushing whole islands underwater (darkness)
10. Millions of environmental refugees fueling unrest (killing of firstborn)

Ten Things To Do To Help Avoid These Plagues

1. Drive less (buses, trains, bikes, walking!)
2. Drive smarter (better mileage car, fully-inflated tires, tune-ups)
3. Turn your thermostat down (in winter) and up (in summer)
4. Turn off lights and unplug appliances when not in use
5. Install timers and light-detectors for your lighting needs
6. Buy Energy Star, ultra-efficient appliances and electronics
7. Switch to green power where possible; install wind & solar units
8. Organize!
9. Educate!
10. Vote your conscience; vote for the future

What Is a Sustainable Pesah?

Pesah is the time of year most associated with Jewish tradition's emphasis on eating seasonally. Also known as "Hag Ha-Matzot" (possibly a holiday celebrating the new barley harvest) and "Hag Ha-Aviv" (holiday of spring), Pesah is a time to notice and celebrate the coming of spring. The Seder plate abounds with seasonal symbols: the roasted lamb bone celebrates lambs born in spring; *karpas* symbolizes the first green sprouts peaking out of the thawed ground; and a roasted egg recalls fertility and rebirth.

Pesah offers a perfect opportunity to combine the wisdom of a traditional Jewish holiday with our contemporary desire to live healthily and sustainably in our world. Even if you don't normally think about sustainability, we encourage you to try it this Pesah.

What Can I Do?

1. Unplug appliances, and turn off all your lights and desktop computers if you will be gone during the break.

2. Shut your windows and keep the blinds closed to reduce energy consumption.

3. Use eco-cleaning products from companies such as Seventh Generation or Ecover.

4. Pesah gives us a chance to take a break from processed foods that contain refined grains and high fructose corn syrup. Go to your local farmers market and check out your supermarket's selection of organic produce, and enjoy fresh, whole, nutritious fruits and veggies!

5. Celebrate Earth Day. In 2009 the United Nations designated April 22 International Mother Earth Day.

Hametz-Free Seder Ideas

1. Take an "environmental inventory" of the chemicals in your home and synagogue as you clean out hametz, and determine whether any of them should also be removed in the proper manner.

2. Make your Seder plate sustainable! *Karpas*: Grow your own or buy at the local farmers market. *Beitzah*: Buy organic or free-range eggs. *Maror*: Buy a whole root, or try alternative bitter kosher le-Pesah products. *Haroset*: Buy local ingredients. Pesah offering: For a cruelty-free alternative, try a roasted beet!

3. Discuss whether or not the Jewish concept of freedom extends to rights for animals, plants, and even ecosystems. How far does freedom go? Is "eco-justice" compatible with Jewish tradition?

4. "Observe the month of Aviv and make the Pesah offering to the Lord your God who brought you out of Egypt" (Deut. 16:1). Why must Pesah be observed in the spring? Make this an additional question at your Seder.

5. Add to your Seder readings from Jewish sources such as "Judaism Eternal," (Samson Raphael Hirsch), "The Rhythms of Jewish Living" (Marc Angel), or "God in Search of Man" (Abraham Joshua Heschel), which serve to heighten awareness of the natural phenomena that return each spring to remind us of the oneness of the God of freedom and the God of nature.

6. Take a walk in your community and think about what Solomon might have been thinking as he "went down to the nut garden to look at the fresh plants by the stream, to see whether the vine has blossomed and if the pomegranates are in flower" (Song of Songs 6:11). What would you look for in such a walk in your own community?

Third, Canfei Nesharim reminds us that on Sukkot, we celebrate water through an ancient ritual called the "Simchat Beit Hashoevah" (Celebration of the Water Drawing Ceremony) and we continue recognizing the value of water through "tefilat geshem", the beginning of our prayers for rain. Our rabbis and ancestors understood that water is essential for life. Each year, Canfei Nesharim encourages communities to focus on appreciating water on Sukkot and Shemini Atzeret.

The following is a program Canfei Nesharim invites communities to run each year, in celebration of water.

True Joy for your Simchat Beit Hashoevah

What to Provide:
- Chocolate and other delicious treats
- Cakes, crackers, or cookies (falling into the Jewish food category of 'Mezonot'). When you eat these items, you are considered to be "dwelling in the Sukkah" and can make the special blessing for this: Barukh... asher kidishanu b'mitzvotav v'tzivanu leshev basukkah.
- Water to drink (the water should be served in pitchers rather than in bottles – if you must buy bottles, buy 3 gallon jugs of water; serve in paper cups, not plastic)

After the group has settled in and had the chance to eat, quiet the group for a brief discussion. It is ideal if they can sit.

Share the following with your audience:
Talmud, Tractate Sukkah 51 a+b
He who has not seen the rejoicing at the place of the water-drawing has never seen rejoicing in his life.

What made for such wonderful Simcha (rejoicing) at the Simchat Beit Hashoevah?

Rambam, Laws of Shofar, Sukkah and Lulav 8:13
And how was this Simcha [expressed]? The musicians play flutes, harps, fiddles etc. And every person plays the instrument that they know how to play. And whoever knows how to sing, sings. And each person dances and claps and dances and parties the way they know. And words of song and praise are said...

Talmud, Tractate Sukkah 53a
It was taught: They said of R. Simeon b. Gamaliel that when he rejoiced at the Rejoicing at the place of the Water-Drawing, he used to take eight lighted torches [and throw them in the air] and catch one and throw one and they did not touch one another; and when he prostrated himself, he used to dig his two thumbs in the ground, bend down, kiss the ground, and draw himself up again, a feat which no other man could do, and this is what is meant by Kidah [Hebrew for bowing in appreciation and respect].

The celebration of the Simchat Beit Hashoevah was a celebration of water. "On Sukkos we are judged regarding water." (Mishnah Rosh HaShanah 1:2) Here we have provided you with water, along with many other delicious treats. Which treats did you enjoy more, the treats, or the water? You probably enjoyed the treats. Did you even drink the water?

Today, we've become so accustomed to water, that we take it for granted. But the Rabbis understood that water is much more precious than these other treats:

Midrash Genesis Rabbah 13:3-4
Said R. Shimon bar Yohai, 'Three things are of equal weight with one another, and these are they: the earth, man, and rain." Said R. Levi bar Hiyyatah, 'And each of the three of them is written with three consonants, to teach you that if there is no earth, there can be no rain, and if there is no rain, there can be no earth, and if the two of them are not, then there can be no man.'

At this Simchat Beit Hashoevah party, we invite you to appreciate the value of water by taking actions to protect it.

The State of Our Water[162]
- More than half of the world's major rivers are seriously depleted and polluted. In the United States, more than a third (39%) of streams and rivers are impaired by pollution or habitat degradation, and an additional 8% are threatened.
- 1.1 billion people in the world do not have access to safe drinking water and 2.4 billion people lack access to basic sanitation.

- Twenty-five percent of the world's population lives in countries approaching a position of serious water stress.
- Large predatory fish in our oceans have been reduced to a mere 10% of pre-industrial levels. That means that 90% of large fish (including tuna, marlin, sharks, cod, and halibut) have been removed.

What can I do?
- Turn off the faucet! Don't let it run between washing for netilat yadayim [ritual hand-washing before eating bread], while brushing your teeth, or while lathering dishes.
- Eat sustainable seafood.
- In your yard, select plants that have low requirements for water, fertilizers, and pesticides.
- Use non-toxic cleaning alternatives and low-phosphate or phosphate-free detergents.
- Take unwanted household chemicals to hazardous waste collection centers; do not pour them down the drain.
- Eat less beef.
- Don't buy bottled water. If you feel that your drinking water is not safe, filter your water.
- Buy eco-friendly products from companies like Seventh Generation.

For more information and resources to run this program in your community, visit www.canfeinesharim.org/sukkos.

More resources on other holidays are available from Sviva Israel, Canfei Nesharim and the Coalition on the Environment and Jewish Life (COEJL).[163] You can also explore a full library of Jewish environmental resources on a wide range of holidays and topics at www.Jewcology.com.

Put it into Action!

Ways I Can Make My Jewish Holidays More Environmentally Conscious:
Rosh Hashanah/Yom Kippur:
Sukkot/Shemini Atzeret:
Hanukkah:
Tu b'Shevat:
Purim:
Pesah:
Shavuot:
Three Weeks:

Number Thirty-One

Celebrate Tu b'Shevat

Tu b'Shevat, the Jewish "new year of the trees,"[164] has increasingly become a Jewish-environmental holiday in recent decades. In the land of Israel, this day — the fif-

Recommended Action:
- Celebrate Tu b'Shevat with your family or community this year.

teenth of the Jewish month of Shevat —is the time of year when the sap begins to rise in the earliest-blooming trees, thus beginning a new fruit-bearing cycle.

Legally, the day marks the time when tithes are separated from produce grown in the land of Israel. Fruit which grows before the 15th of Shevat counts toward the previous "fiscal year," as it were, whereas fruit which grows on the 15th of Shevat or after counts toward the next fiscal year.

The Kabbalists of Tzfat (Safed) expanded on this Talmudic understanding, developing a "Tu b'Shevat Seder" which focused on the benefits of trees, appreciation for the abundance of fruits and blessings from the Creator, eating many fruits from the land of Israel, and increasing levels of holiness toward connection with God.

Today, the "birthday of the trees" is celebrated in schools and Jewish communities worldwide, with activities, recognition of the blessings trees provide, and an increasing environmental consciousness.

Modern Tu b'Shevat Sedarim often include environmental thoughts alongside ancient mystical and Talmudic texts, the drinking of wine or grape juice and the eating of lots and lots of fruit and nuts. It's a fun and engaging activity that can be tailored for children, adults, and families, and can be both playful and mystical as the level of the audience dictates.

A wide range of Jewish environmental organizations offer free activities and resources to help families and communities celebrate Tu b'Shevat with a sense of environmental awareness and rededication. This year, find out if Tu b'Shevat activities will happen in your community, and if not, organize one yourself. Great resources exist on the following websites:

- Jewish National Fund: Planting trees in Israel is the way to observe Tu b'Shevat! Also JNF has a wide range of sermons and programmatic resources for Tu b'Shevat. http://support.jnf.org/site/PageServer?pagename=treesource
- http://jnf.org/trees
- Jewcology.com – a "one-stop shop" of Jewish-environmental resources shared by a wide range of Jewish environmental organizations, with a robust library of Tu b'Shevat materials.
- The Coalition on the Environment and Jewish Life (COEJL) offers numerous Tu b'Shevat program ideas at http://www.coejl.org/~coejlor/tubshvat/celebrate/.
- Hazon offers a printable Tu b'Shevat Seder manual at http://www.hazon.org/food/tuBishvat/Seder_Manual.pdf.
- Canfei Nesharim: A wide range of Torah learning and Tu b'Shevat program resources, including Torah study sheets, children's activities, colorful one-page haggadot and eco-reminders (stickers, magnets, keychains) for order: http://canfeinesharim.org/community/shevat.php
- Jewish Reconstructionist Federation: Great list of Tu b'Shevat resources across the Jewish-environmental community: http://jrf.org/Tu-B-Shevat

Put it into Action!

Tu b'Shevat

What's happening in my community this year?

What could I help create?

Number Thirty-Two

Plant and Protect Trees

"As the days of a tree so shall be the days of my people"
Isaiah 65:22

"Rabbi Yohanan ben Zakkai said: "If you have a sapling in your hand, and someone says to you that the Messiah has come, stay and complete the planting, and only then go to greet the Messiah."
Avot de Rabbi Natan 31b

"It is forbidden to live in a town which has no garden or greenery"
Talmud, Kiddushin 66a

> **Recommended Actions:**
> - Learn about local trees in your area. Where do trees need to be planted or protected?
> - Plant trees or other plants where they would be helpful.
> - Buy FSC-certified and 100% PCW recycled wood products, to ensure that ancient forests are being protected.

One day Honi the circle-drawer was journeying on the road and he saw a man planting a carob tree. He asked him "How long does it take for this tree to bear fruit?"
The man replied, "Seventy years."
Honi further asked him, "Are you certain that you will live another seventy years?"
The man replied, "I found grown carob trees in the world; as my ancestors planted these for me, so I too plant these for my children."
Talmud, Ta-anit 23a

By planting trees we help keep our environment stable and safe. "If not for the trees, human life could not exist." *Midrash Sifre* (on Deuteronomy 20:19)

Jews have been tree-planters since ancient times. With the birth of the modern Zionist movement, reforesting the Land of Israel became an important way to advance the re-establishment of a Jewish state.

On the web site of the Jewish National Fund (www.JNF.org) we find this description of the international Jewish effort to reforest Eretz Yisrael:

Israel is one of only two countries in the world that entered the 21st century with a net gain in its number of trees. But Israel was not blessed with natural forests; its forests are all hand-planted. When the pioneers of the State arrived, they were greeted by barren land. To claim the land that had been purchased with the coins collected in JNF blue and white pushkes, the next order of business was to plant trees among the rocky hillsides and sandy soil.

Since it was established in 1901, JNF has planted more than 250 million trees all over the State of Israel, providing luscious belts of green covering more than 250,000 acres. JNF national forest development work creates "green lungs" around congested towns and cities, and provides recreation and respite for all Israelis. While the forests of Israel belong to the people, JNF ensures their environmental soundness and is focusing on diversification, planting trees indigenous to the Middle East such as native oaks, carob, redbud, almond, pear, hawthorn, cypress and the exotic Atlantic cedar. (www.jnf.org/trees)

Let's focus for a few moments on the benefits of planting trees in our neighborhood, and on the grounds of our Jewish communal institutions (synagogues, supplemental and all-day religious schools, JCCs, Federations, etc.).

Trees are very important in maintaining the earth's environmental cycle. They provide habitat and food for many species of wildlife. Long-lived trees also accumulate carbon and take it out of the atmosphere. This reduces the risk of global climate change, because the CO_2 is stored in the trees, instead of in the air.

By planting trees and shrubs around our places of habitation and work we lower surrounding air temperatures during warm months, and reduce our wall and roof temperatures by 20-45°F, keeping our buildings naturally cooler.[165] Trees also protect against erosion.

Planting trees is good exercise, and costs little or nothing. The Arbor Day Foundation offers ten free trees when one purchases an annual membership ($15.). Their web site provides wonderful information on planting, growing and much more about trees and their benefits – www.arborday.org. You can also plant trees with Plant a Tree USA at http://www.plantatreeusa.com. Of course, you can also plant trees in Israel with the Jewish National Fund at www.jnf.org

Each of these organizations will provide beautiful certificates that are sent to those who are honored with the tree planting. Many people who are celebrating a simcha, such as a bar/bat mitzvah, wedding, or any lifecycle milestone, have chosen to distribute such certificates to their guests as economical, attractive and useful gifts to their guests. Some have even used the certificates as invitations to their party.

In addition to planting trees, it's very important to protect existing forests, including old-growth forests and tropical rainforests, such as the Amazon, which is often referred to as "the earth's lungs." You can protect forests by buying products that are certified by the Forest Stewardship Council (FSC). Buying FSC-certified tree products ensures that you are avoiding illegally logged forests and that you are supporting companies that maintain the long-term social and economic well-being of forest workers and local communities, and maintain forests of high value. In addition to compliance with all of the above, plantations must contribute to reduce the pressures on and promote the restoration and conservation of natural forests.

Finally, the best way to avoid using trees when you don't need to is to buy 100% post-consumer waste (PCW) recycled paper. This paper was made out of old newspapers or cardboard boxes that have been reused. When you buy 100% PCW paper, no more trees need to be cut down for your use.

Let's appreciate trees and all they provide to us, by protecting existing trees and planting new ones.

Put it into Action

Local Trees I Can Care For:

Where can I plant a tree?

Other actions:

Section J: Build a Green Jewish Community

"One generation goes and another comes, but the earth remains forever." Kohelet 1:4

The Jewish community has an opportunity to be a leader in the environmental challenges we face today. Environmental choices can often save money and improve our health.

Many young Jews are excited about the environmental movement, and are looking to the Jewish community to show the way. By taking on environmental actions within the context of the Jewish heritage, we can inspire young Jews who are committed to the environment. Some have suggested that the number one project for the Jewish community, besides helping to promote the welfare of the State of Israel, is the environmental movement, just as the Civil Rights, anti-Vietnam War and Soviet Jewry movements were in the sixties and seventies.

By preserving resources such as energy and water, we can also help our brothers and sisters in the land of Israel. Water resources are scarce in Israel, and we demonstrate our concern by recognizing the preciousness of this resource. Saving energy and supporting renewable energy also can help address the geopolitical imbalances that have resulted from our reliance on oil.

Yet the most important reason to build a green Jewish community is because it's just the right thing to do, for ourselves, our children and future generations. Jews have often been at the forefront of critical social movements such as this one. Let's take a leadership role and demonstrate the relevance of our ancient heritage of an issue of great importance today.

SECTION J: BUILD A GREEN JEWISH COMMUNITY

Number Thirty-Three

Get to Know the Jewish Environmental Movement

If you think ahead one year, plant a seed.
If you think ahead 10 years, plant a tree.
If you think ahead 100 years, educate the people.

Ancient Chinese proverb

> **Recommended Actions:**
> - Get to know the broad range of Jewish environmental organizations in the United States.
> - Use their resources to help learn about Judaism and the environment, and to teach these ideas in your community.

Numerous Jewish environmental organizations now offer a range of opportunities for you to learn about the Jewish-environmental connection and develop skills to engage others in your community. Here are a few environmental organizations and their campaigns.

Jewish National Fund, www.jnf.org, (New York, NY) the oldest and largest Jewish environmental organization provides 1200 Jewish schools with free educational materials on Israel and the environment; provides free World Water Monitoring Day kits to schools in the United States and Israel funded by the U.S. Forest Service; builds rainwater harvesting programs and provides educational programs to schools in Israel.

Hazon, www.hazon.org: (New York, NY)
- Helps local Jewish communities begin "Community Supported Agriculture" (CSA) programs with a Jewish communal twist.
- Organizes the "Jewish Climate Change Campaign," including a pledge you can sign and distribute to encourage a more sustainable Jewish community.
- Organizes regular "Jewish Environmental Bike Rides" across the United States and in Israel, where you can learn more and connect with like-minded Jewish environmental activists.

Coalition on the Environment and Jewish Life (COEJL): www.coejl.org (New York, NY)
- Offers a wealth of programmatic suggestions for communities on their website.
- Organizes a Jewish Energy Covenant Campaign to engage the Jewish community in more sustainable energy use, including a CFL campaign each Hanukkah.

Canfei Nesharim: Sustainable Living Inspired by Torah: www.canfeinesharim.org (Silver Spring, MD)
- Offers a Torah teaching about the importance of the environment for each Torah portion of the year, including source sheets.
- Offers Torah teachings, study guides, and program suggestions for each Jewish holiday.
- Partners with synagogues and schools to implement programs that introduce environmental sustainability.

Teva Learning Alliance: (http://tevalearningalliance.org/) (New York, NY)
- Runs a camp for Jewish elementary school students to connect to Jewish environmental wisdom.
- Offers fellowships for young people to work at the camp and have an in-depth Jewish environmental experience
- Organizes an annual seminar for Jewish educators

Wilderness Torah, www.wildernesstorah.org (San Francisco area)
- Offers experiential land-based festivals, including Sukkot on the Farm, Pesah in the Desert, Shavuot on the Mountain and Tu B'shvat in the Redwoods.
- Jewish Vision Quest and Wilderness Walks for adults, and a B'nai Mitzvah Rite of Passage Program for youth.
- Gan Torah, weekly outdoor experience for 8-10 year olds that teaches children to learn and experience Torah and Jewish life in tune with nature and seasonal cycles.

Jewish Farm School (www.jewishfarmschool.org) (Philadelphia, PA)
- Offers Organic Farming Alternative Breaks.
- The Anafim internship, for high school seniors to gain practical skills, leadership training, and engage in thought provoking study.

Kayam Farm at Pearlstone (http://www.pearlstonecenter.org/kayam.html) (Baltimore, MD)
- Offers an annual Beit Midrash program, hands-on farming projects, and an 8-week summer Kollel program including Torah learning and farming.

Eden Village Camp (http://edenvillagecamp.org/) (Putnam Valley, NY)
- A non-profit Jewish environmental overnight camp for 3rd – 11th graders.

The different Jewish environmental organizations also collaborate on several common projects intended to engage the entire Jewish community:
- www.Jewcology.com, a website which will enable you to search for a wide range of Jewish environmental resources for your community.
- Shabbat Noah: Annual Program focusing on climate change, energy and sustainability, focusing on Shabbat of the Torah portion of Noah.
- Tu b'Shevat: a wide range of resources are available.
- Jewish Climate Change Campaign: a wide range of Jewish environmental organizations are collaborating to address climate change.

To get to know the Jewish environmental movement, find a Jewish environmental organization or project that speaks to your values. Then join, volunteer, donate, and get involved!

You can also learn about the environment and sustainability from courses in your neighborhood or online. Take a look at your local community college, adult education organizations such as "Communiversity," Elderhostels, local synagogues, Hadassah chapters and other organizations. If your synagogue does not offer such a course, speak to the rabbi or adult education chairperson and encourage them to inaugurate such a course.

Put it into Action!

Jewish Environmental Organizations I'd Like to Join:
___
___
___

Resources I Can Use to Educate My Community:
___
___
___
___

Number Thirty-Four

Help Your Synagogue and Other Jewish Institutions Go Green

Our ancestors acclaimed the God
Whose handiwork they read
In the mysterious heavens above,
And in the varied scene of earth below,
In the orderly march of days and nights,

Lift your eyes, look up;
Who made these stars?
God is the mystery of life,
Enkindling inert matter
With inner drive and purpose.

Rabbi Mordecai M. Kaplan

> **Recommended Actions:**
> - Form a green committee in your synagogue or other Jewish institution.
> - Look for opportunities to help your community "go green."

There are many things that synagogues, JCCs, Federations, Jewish Family Services and other Jewish organizations can do to green their buildings, their programs and their entire institution.

Every Jewish organization should have a "Green Committee." This green committee should make regular reports to the Board of Directors, and to the membership through the institution's publications about the projects it is involved in, in helping make the group more environmentally friendly. Whether it be the physical facility, the food that is served, the programs that are offered, the education that is planned, etc., the constituency should be constantly brought up to date on what is happening, and what future plans are being prepared to forward the goals of the committee. The committee should have a clear mission statement, and work towards doing its share to make the institution more "green."

Some excellent partners for a community green project include:
- Interfaith Power and Light (chapters nationwide)
- GreenFaith (New Jersey)
- Jewish Greening Fellowship (New York region)
- Canfei Nesharim (leadership training for lay leaders, nationwide)

"Greening Synagogue Resources," compiled by Rabbi Fred Scherlinder Dobb for the Coalition on the Environment and Jewish Life (COEJL), recommends seven different areas where we can make a different in our synagogue community. Sample actions are listed here, but each area includes a link to additional resources. For the entire set of resources, visit http://www.coejl.org/~coejlor/greensyn/gstoc.php.

1. BUILDINGS

Get Energy Star programmable thermostats, divided by zones, so you're not heating or cooling the building beyond what's necessary. Cutting back on the heat or A/C by just 1 degree saves an average of 3% on your utility bill -- and on your greenhouse emissions.

2. GROUNDS

Plant native species around your building, which provide much-needed habitat for local birds and other creatures while also needing less water and no chemicals.

3. PURCHASING

Reduce, reuse and recycle in the office: print fewer copies than needed and let people share them; keep a pile of clean-on-one-side paper for use in printers & copy machines; recycle used paper; and purchase paper with high post-consumer recycled content.

4. PROGRAMS

With your social action committee or other group within the synagogue, plan events that are social, educational and tikkuning-the-olam all at once – like Torah-nature hikes while picking up trash, or pulling non-native weeds from nearby woods.

5. YOUTH EDUCATION

Implement at least one of the many great curricula that teach our young people about nature and Judaism together – kids are ripe for it and the materials are out there.

6. ADULT EDUCATION

Teach a timely topic that conveys Creation care together with Torah teachings -- such as the shiurim (text studies) on Jewish responses to global climate change and biodiversity.

7. RABBINIC

For rabbis, take advantage of the sermon-starters and notes on integrating environmental concern into life-cycle events {available on COEJL's website, see link above}. For non-rabbis, feel free to do the same – and to tell your rabbi about these resources!

Recently I drove to the local health food store in my hometown, Princeton, NJ, called "Whole Earth Center," and noticed that in the parking lot there were designated parking spaces that were reserved for hybrid cars. Like all of us, I have seen in almost every parking lot I ever saw, special reserved places for handicapped persons. I had never seen reserved parking for hybrid cars. This sends an important signal to the community: that buying a hybrid car is a good thing, and will bring you a special place of honor.

The idea occurred to me that synagogues and other Jewish institutions that have parking lots might consider doing the same thing. What a powerful message, what a great educational it would be, for every member of a synagogue, or JCC, or Federation, etc. to see in the parking lot a proud sign: "Reserved for Hybrid Cars." {See above, chapter 7, "Green Your Transportation," for more on this}.

Synagogues can also partner with a town, synagogue, school, or even a university in Israel, to make their local "world" more green.

If your community doesn't have a Green Committee yet, start one!

"We are one.... Together we suffer, together we exist, and forever we will recreate each other." Pierre Teilhard de Chardin

Put it into Action!

Who could form a Green Committee with Me?

What will I need to get started?

Actions I Will Take:

Number Thirty-Five

Serving Healthy Food in Jewish Institutions

Food is a key component in every Jewish activity, and nearly every time we walk into a Jewish institution we're presented with food. Whether that food is healthy and sustainable is up to us.

> **Recommended Action:**
> - Encourage your synagogue and other Jewish institutions to serve more healthy and sustainable food.

Reciting the prayers before (Ha-motzee) and after (Birkat HaMazon) meals helps to place the act of consuming food in its proper context, acknowledging God as the Source of all provisions.

GreenFaith (www.greenfaith.org) offers detailed information about choosing healthy and sustainable food for religious events. The information is available in a downloadable booklet called "Repairing Eden." For a copy, write to info@greenfaith.org. Here we will briefly summarize the recommendations from the booklet.

When choosing food for your Jewish community event, consider the following ways to make it more healthy and sustainable:

- **Use Fair Trade, Shade Grown, Organic Beverages:** Provide fair-trade, shade-grown, organic tea and coffee, which are grown and processed in an environmentally and socially responsible manner, treating workers and the environment with respect and care. If available, purchase products that are Rainforest Alliance Certified and/or Bird Friendly Certified.
- **Support Local Bakeries**: reducing the amount of energy required to bring baked goods to your plates. Also, see if you can find sweets that use local ingredients.
- **Choose Healthy Ingredients:** Buy snacks that are lower in fat, and choose food that is local and/or organic (see food section for more info on these topics). Also, choose products that are made without trans-fats (also known as hydrogenated or partially hydrogenated oils).

- **Consider fruits and vegetables** as possible refreshments, and buy them local and/or organic if possible.
- **Serve Vegetarian and Vegan Meals:** from time to time, make a meal, Kiddush, or Oneg Shabbat entirely vegetarian. And to support those who are making more sustainable choices, always provide a vegetarian/vegan option at community meals. The vegetarian/vegan selection should be a full meal that changes with similar frequency as the meat options. For instance, a side salad, particularly if the toppings do not include protein, or steamed vegetables, are not full vegetarian meal options.
- **If you must serve meat, consider options for more sustainable meat choices.** More and more farms are raising animals in human conditions that are also better for the planet. Sustainable kosher meat is now available in many communities across the United States. For example, KOL Foods (www.KolFoods.com) provides kosher grass-fed beef, lamb, and poultry and can ship across the United States. Another example is Grow and Behold, (www.GrowandBehold.com). Other local options may be available to you as well.
- **If you will be serving seafood, choose fish that is managed sustainably.** The most up-to-date resource on sustainable seafood is Monterey Bay Aquarium's Seafood Watch, which can be viewed at http://www.montereybayaquarium.org/cr/seafoodwatch.aspx. The guide is continually updated, and offers sustainable fish options both regionally and nationally. In addition to following the guidelines in Seafood Watch, aim to purchase seafood with the Marine Stewardship Council Certification label.
- **Use Whole Grains,** such as those used in whole wheat bread, which are much higher in vitamins and fiber than processed grains. Purchase products that contain the Whole Grain label or read the ingredients and look for the word "whole" preceding the grains listed. Offer whole and unrefined grain as a side dish; choose brown rice over white rice, or offer millet, couscous, barley, buckwheat or quinoa.
- **If members of the community will contribute to the food served, encourage them to consider these standards as well.**

To find local, fresh, sustainable food, enter your zip code at: www.eatwell-guide.org.

Put it into Action!

How Could We Improve the Food We Serve in My Community?

Who can I talk to about this?

Number Thirty-Six

Learn about and Protect Israel's Environment

Israel has developed a fascinating and important set of environmental innovations, from drip-irrigation that allows for agriculture in a water-scarce climate, to the increase of trees promoted and planted by the Jewish National Fund. JNF has also built over 210 reservoirs filled with collected rain water and recycled water and 40% of all water used by farmers is recycled water. Israel leads the world with the highest percentage of its water reused—77% in 2011! Israeli homes are required by law to have solar water heaters, and an initiative is underway to pilot the use of electric cars.[166] When you travel to Israel, you probably notice all kinds of water and energy-saving devices and practices. Indeed, we should be quite proud of these Israeli environmental innovations and the wisdom of "limited resources" that they represent.

Recommended Actions:
- Learn about Israel's environmental challenges, and find out what you can do to address them.
- Support organizations that are working for sustainability in the land of Israel.

On the other hand, Israel still has many environmental challenges. Jews around the world should be aware of Israel's environmental issues, and do whatever we can to ameliorate these problems. The list below, of environmental problems in Israel, appears on the web site of the Green Zionist Alliance, www.greenzionism.org:

Israel's Environmental Challenges

By the Green Zionist Alliance: The Grassroots Campaign for a Sustainable Israel

"Israel's Environmental Challenges" Copyright the Green Zionist Alliance. Used with permission. For more information on Israel's environment, please visit: www.GreenZionism.org

SECTION J: BUILD A GREEN JEWISH COMMUNITY

AIR POLLUTION: Israel's major cities — Jerusalem, Tel Aviv and Haifa — as well as industrial centers like Ashdod, face severe air-pollution problems, primarily from industrial and automobile emissions. In 2003, the Israel Union for Environmental Defense (IUED) published the results of a study it conducted with the U.S. Environmental Protection Agency that indicated that 1,400 Israelis die each year from exposure to air pollution in Tel Aviv and Ashdod alone. This is more than twice the number of Israelis who die annually due to traffic accidents and terrorist acts combined. One out of six Israeli children suffers from asthma, primarily caused by air pollution.

WATER POLLUTION: All of Israel's rivers, except those flowing through sparsely populated areas, are much more polluted than rivers in Europe and the United States. For example, the Kishon River has been especially hard hit because for more than 40 years Haifa Bay's chemical industry discharged its raw industrial wastes directly into the river. Israel ranks 88th out of 122 selected countries in terms of water quality, according to the 2003 United Nations World Water Development Report. Water quality is higher in several Third World countries. A recent nationwide survey found that more than half of Israel's drinking-water wells are significantly polluted.

WATER SHORTAGES: Potentially severe water shortages may become the most crucial problem that Israel will face, touching on its very existence. Since the mid-1970s, demand for water has often times outstripped supply. Israel is a semi-arid country where no rain falls for at least six months a year. While Israel was known as a country that practiced water conservation and pioneered the development of the drip irrigation method, the country recently has been using increasing amounts of water per person, often for non-essential uses. Israel's main water sources are expected to continue to decline, endangering drinking-water quality, and raising the specter that it will soon not be possible to supply sufficient drinking water.

POPULATION DENSITY AND LOSS OF OPEN SPCE: Israel is one of the world's most densely populated countries. With more than 6.5 million people in a country about the size of New Jersey, Israel is more densely populated than India. Safeguarding its precious land resources is a major national challenge. This will be more critical as Israel's population continues to grow. Poor planning and improper development is leading to accelerated suburbanization in Israel's densely populated central region. Economic pressures for urban development are leaving towns bereft of parks, gardens and play areas that are essential to health and quality of life.

WASTE DISPOSAL: Israel faces a solid-waste crisis due to increasing amounts of garbage and the country's meager land resources. Solid-waste disposal in Israel causes significant and irreversible damage to Israel's groundwater, air, soil and quality of life. In 2000, then-Israeli Environmental Minister Dalia Itzik stated that she regarded garbage disposal as Israel's number-one environmental problem.

SPECIES EXTINCTION: Pollution, tourism and urban sprawl all lead to this pressing environmental concern. Many corals in the reefs of Eilat are threatened with local extinction, migrating birds find less land on which to forage annually along their migrations, and numerous animal species are threatened as their habitats are replaced by housing, highways and other development.

DEATH OF THE DEAD SEA: One significant result of the water shortages in Israel is that the Dead Sea is shrinking rapidly. The Dead Sea relies on the fresh water of the Jordan River, and that once-wide river has become just a contaminated trickle, as water has been diverted for agricultural and other uses. As the sea's water disappears, large sinkholes are created that make it dangerous to be near the sea in some areas.

All Jews should be concerned with the environmental state of the land of Israel, and support efforts and policy initiatives that ensure the long-term sustainability of resources in this precious Jewish homeland. Some things you can do include:
- Learn about Israel's environmental challenges and work to address them.
- Support environmental organizations in Israel, such as the Jewish National Fund, Society for Protection of Nature in Israel (SPNI) and Adam Teva v'Din, the Israel Union for Environmental Defense. A great list of Israeli environmental organizations can be found at http://www.coejl.org/resources/israelorg.php.

- Support the Green Zionist Alliance (GZA), a new party promoting environmental policies in the World Zionist Organization (WZO).
- Teach others about the environmental ingenuity and the environmental challenges faced by the nation of Israel.

Additional articles are in Figure 13.

Put it into Action!

Israel's Environment
Things to be proud of:
Things to make better:
What can I do?

Figure 13

Articles about Israel's Environmental Innovations

Israeli Campus Goes Green, Cuts Costs

February 10, 2010

Israel21c[167]

Israel's University of Haifa has cut its electricity consumption by 22 percent, fuel oil consumption by 64% and water consumption by 11%, reducing costs and minimizing CO_2 emissions and air pollution.

Last year the university was certified as a "Green Campus" by the Ministry of Environment and its impressive achievements in protecting the environment since October 2008 also include an 8% decrease in paper use, a 100% increase in paper recycling and the removal of 15 tons of electronic waste from the campus for safe disposal.

Most of the energy savings resulted from a number of projects carried out by the university's Building and Maintenance Division, headed by the university engineer, Nabi Amar: Old cooling units were replaced; a more efficient heating tank was installed; old light bulbs were replaced with energy-saving bulbs; air conditioning operation was streamlined; and water-saving toilet tanks and faucets were installed. In addition, new flora to be planted will be native Israeli flora that requires minimal irrigation.

The university has also adopted the nearby Neder River in the Carmel as a project to benefit both the environment and the community, and has held educational and clean-up activities on its banks.

UN: Israel #1 World Leader in Water Recycling
March 23, 2009
Israel National News[168]

by Malkah Fleisher

Despite Israel's constant struggle to maintain a sufficient water supply — or perhaps because of it — Israel was named the world's most efficient recycled water user in a United Nations report issued in honor of International Water Day.

Presented at the 5th World Water Forum in Istanbul, the UN report also ranked Israel as one of the world's leaders in desalinated water use.

Officials at the Israel Export and International Cooperation Institute noted that water technology exports have doubled since 2005, with 200 Israeli companies exporting $1.4 billion worth of water management, recycling and purification, irrigation, desalination, and safety technologies to over 100 countries in 2008.

Israel purifies and reuses almost 70 percent of its waste water each year for agriculture. Much of the leftover sewage water is reused for other purposes.

Israel vastly outranks even the second most efficient recycled water user — Spain — which only recycles 12 percent of its waste water for agriculture. In Turkey, the host country of the forum, only 3.6 percent of sewage water is recycled.

According to the UN, approximately 1.4 million children die annually as a result of drinking polluted water. In China, roughly 10,000 chemical plants are located on the banks of the country's two largest rivers.

Environmentalist group Friends of the Earth-Middle East, recommends that Israel set its next goal at reducing domestic water consumption, which is still significantly higher than in other developed countries.

Tel Aviv is Going Green
AIPAC Update, 4-16-2010 - www.aipac.org

Tel Aviv is setting new building regulations based on the principles of so-called green construction, *Haaretz* reported Tuesday. Under the new rules, building permits will be handed out if the engineering and architectural plans are environmentally friendly, and will be issued by the city engineer only after receiving an expert opinion on the environmental aspects of the plan. Requirements for green construction include lower water use, energy efficiency, thermal and acoustic insulation, limits on the use of raw materials, natural ventilation and prevention of mold, natural lighting, so-called green roofs and recycling. Green construction also uses advanced technologies in order to save energy.

Concluson

Become a Jewish-Environmental Activist

"If one Jew sins, all of Israel feels it….This can be compared to the case of men on a ship, one of whom took a drill and began drilling beneath his own place. His fellow travelers to said to him: 'What are you doing?' He replied: 'What does that matter to you; I am drilling only under my own place?' They continued: 'We care because the water will come up and flood the ship for us all.'" *Midrash Vayikra Rabbah* 4:6

"Always remember, you have within you the strength, the patience, and the passion to reach for the stars to change the world." Harriet Tubman

After starting to implement many of the ideas in this book, and making your home and your life more green, you will begin to feel that there is much more to do in the world to make our planet cleaner, healthier, and safer for the next generation. At some point you will perhaps want to become more involved in the Jewish environmental movement.

SECTION J: BUILD A GREEN JEWISH COMMUNITY

Jewish environmentalism offers something deeper than just being part of the environmental movement. It offers us the opportunity to connect our Jewish identities, and the deep wisdom of our ancient Jewish tradition, to the modern moral and ecological challenges we face today. Jewish perspectives on the sustainability of our planet can enrich our environmental understanding, and can also help us think about long-term needs and choices that will enable us to bequeath a healthy planet to future generations.

Within the Jewish community, we can deepen our understanding of the environmental challenges we face, possible solutions, and the Jewish ideas that can help us think differently about protecting our resources. By doing this, we can make the Jewish community a part of today's environmental solutions, while also connecting more deeply with our rich Jewish heritage.

Here are some ways in which you can become a "Jewish Environmental Activist":
- Join a Jewish environmental organization.
- Explore resources and sign up to participate at www.Jewcology.com, a social networking website for Jewish environmentalists to interact and share ideas.
- Subscribe to Jewish environmental e-newsletters, such as Canfei Nesharim's weekly environmental teaching about the Torah portion, or e-newsletters from COEJL, Hazon or the Teva Learning Alliance.
- Consider contributions to Jewish environmental organizations as part of your charity giving. Include bequests in your will.
- Think about a career in Jewish environmentalism!

To stay informed on environmental topics globally and locally:
- Subscribe to a green magazine or e-zine. One good one is Grist Magazine, www.grist.org. Another is www.treehugger.com.
- Monitor environmentally-focused legislation in your community, state and nation. See www.speakout.com – helps understand legislative process and political activism.
- If you are considering a move to another city, check out the air quality. Cities vary widely in their fuel exhaust pollution. The EPA maintains an Air Quality Index (www.airnow.gov) to evaluate

the air in different locations. *The Green Guide Top 10 Green Cities* can help you make an intelligent and conscious choice. See <u>www.thegreenguide.com</u>.

"We do not inherit the earth from our ancestors, we borrow it from our children," warned Chief Seattle, a well-known 19th century Native-American, who was an early pioneer in promoting ecological concerns.

Let's each of us do our share to protect the environment and give back to our children a world worthy of us and of them.

Rev. Bob Edgar, former Pennsylvania Congressman, and former head of the National Council of Churches, presently president of Common Cause, recently wrote:

"God is calling us to this moment as the Old Testament prophets were called. We have been to the mountaintop and discovered the glaciers melting. We have been to the valley of the shadow of death and seen our fish floating poisoned on the water. The prophetic voice we must hear is not the one that welcomes the squandering of God's creation. It is the still, small prophetic voice calling each of us to do our part to steward and safeguard the planet." (*Middle Church*, pp. 61-2)

"I have set before you life or death, blessing or curse. Choose life, therefore, that you and your descendants may live." Deuteronomy 30:19

Acknowledgments

I am grateful to many people for guiding and advising me during the writing and editing process. My step-sons Yoni and Pesach Jeremy Stadlin made useful suggestions. My gratitude to them for their initial inspiration. My involvement in the green movement because of their influence is too great to be expressed in words.

For specific suggestions in enhancing the quality and accuracy of the book, there are several people who provided expert professional advice. I want to especially acknowledge Evonne Marzouk, Executive Director, and Daniel Weber, Ph.D., chair of Science and Technology Advisory Board, of Canfei Nesharim, for their assistance in reviewing the book for Jewish and scientific content. Dr. Daniel Weber is Senior Scientist at UW-Milwaukee Children's Environmental Health Sciences Core Center. Canfei Nesharim is an organization that educates the Jewish community on sustainable living inspired by Jewish values.

Rabbi David Saperstein, an old and cherished friend and colleague, has made many important suggestions which were incorporated into the text. I am extremely grateful to David for his eloquent Foreword, which he wrote during an exceptionally busy schedule.

Rabbi Eric Lankin, D.Min., Chief, Institutional Advancement and Education, Jewish National Fund, was also very helpful with his many valuable suggestions.

Rabbi Shaul Yudelman, of Jerusalem, was one of the first to read the full manuscript and offer suggestions. I am indebted to him for his knowledge, expertise and commitment to environmentalism and its connection to the Jewish heritage.

I am also grateful to Sybil Sanchez, Director, COEJL - Coalition on the Environment and Jewish Life, and Jeremy Benstein, of the Heschel Center for Environmental Learning and Leadership, Israel, for their constructive comments. My son Jonathan Elkins, of Tel Aviv, helped with his skillful editing. David Krantz of the Green Zionist Alliance provided valuable information to correct a few factual errors.

Any errors, factual or otherwise, are solely the responsibility of the author.

Credits

This page constitutes a continuation of the copyright page. Every effort has been made to trace and acknowledge copyright holders of all the material included in this book. The author apologizes for any errors or omissions that may remain and asks that any omissions be brought to his attention so that they may be corrected in future editions. The author may be reached at DPE@JewishGrowth.org.

While the author is grateful to many web sites for valuable information, he is particularly grateful for material from www.canfeinesharim.org and www.coejl.org.

Grateful acknowledgment is given to the following sources for permission to use material:

Reprinted with permission from Canfei Nesharim: Sustainable Living Inspired by Torah. Additional Torah and science resources are available at www.canfeinesharim.org.

Reprinted with permission from the Coalition on the Environment and Jewish Life – www.coejl.org.

Reprinted with permission from GreenFaith – www.GreenFaith.org.

Reprinted by permission of www.israel21c.org

Reprinted with permission of Rabbi Simkha Y. Weintraub, LCSW, JBFCS' National Center for Jewish Healing. Visit their website at www.ncjh.org.

Reprinted with permission of Rabbi Fred Scherlinder Dobb, Peter Goldberg, Richard H. Schwartz.

Reprinted with permission of Dr. Gabe Goldman, Director of Experiential and Service Learning,

American Jewish University.

Reprinted with permission of Betsy Deutsch (betsy@betsyteutsch.com) and Moti Rieber (moti.rieber@gmail.com).

Reprinted with permission of The Rabbinical Assembly - www.rabbinicalassembly.org.

Reprinted with permission © Cornell Chronicle

"Israel's Environmental Challenges" Copyright the Green Zionist Alliance. Used with permission. For more information on Israel's environment, please visit: www.GreenZionism.org

© Boulder Jewish News, Inc. 2009-11. Used with permission.

Reprinted with permission of Care2.com

Reprinted with permission from www.Rodale.com

Reprinted with permission from www.svivaisrael.org

Appendix

National Jewish Environment Organizations

Augmented from list compiled by Liore Milgrom-Elcott, COEJL Program Manager

American Society for the Protection of Nature in Israel (ASPNI)
Tel: 212.398.6750; 800-411-0966
Fax: 212.398.1665
Email: robin@aspni.org
http://www.aspni.org/; birthright mission: http://israelnature.com/

Canfei Nesharim ("The Wings of Eagles")
Tel: 703-868-5356
Email: info@canfeinesharim.org
http://www.canfeinesharim.org/

Coalition on the Environment and Jewish Life (COEJL)
116 East 27th Street, 10th Floor, New York, NY 10016
(212) 532-7436 | - www.COEJL.org - info@coejl.org

Eden Village Camp
Phone: 877-397-3336
Fax: 877-497-3390
www.EdenVillageCamp.org
392 Dennytown Road
Putnam Valley, NY 10579
Welcome@EdenVillageCamp.org

Green Zionist Alliance
The Grassroots Campaign for a Sustainable Israel
Tel: 347-559-4492
www.greenzionism.org

Hazon
Tel: 212- 644-2332
Email: info@hazon.org
http://www.hazon.org

Isabella Freedman Jewish Retreat Center
800-398-2630
Shir Feinstein-Feit, Creative Director (860) 824-5991 x313
shir@isabellafreedman.org
http://isabellafreedman.org/home

Jewish Farm School
4905 Cedar Ave.
Philadelphia, PA 19143
215-609-4680
E-mail: info@jewishfarmschool.org
www.jewishfarmschool.org

Jewish Global Environmental Network (JGEN)
Tel: 401-863-2499
http://www.jgenisrael.org

Jewish National Fund
Tel: 212.879.9305 x-245; 888.JNF.0099
Fax: 212.570.1673
To plant trees: 800-542-TREE
Email: education@jnf.org
http://www.jnf.org/

Jewish Nature Center
Email: webmaster@njycamps.org
http://www.jewishnaturecenter.org/

National Religious Partnership on the Environment
110 Maryland Avenue, NE - Suite 108
Washington, DC 20002
Phone: (202) 481-6685, ext. 231
Fax: (202) 543-1297
nrpe@nrpe.org
http://www.NRPE.org

New Israel Fund
202-842-0900
202-842-0991 fax
info@nif.org
http://www.newisraelfund.org

Pearlstone Conference and Retreat Center and the Kayam Farm and Environmental Education Center at Pearlstone
P: 410-429-4400
F: 410-429-4723
http://www.pearlstonecenter.org/

The Noah Project
Tel: 020 8123 2859
info@biggreenjewish.org
http://www.BigGreenJewish.org/

The Shalom Center
Tel: (215) 844-8494
Email: office@shalomctr.org
http://www.shalomctr.org/

ShalomVeg.com
Tel: (860) 967-1581
Email: info@shalomveg.com
http://www.shalomveg.com

Teva Learning Alliance
Tel: (212) 807-6376
Fax: (212) 924-5112
Email: teva@tevalearningalliance.org
http://www.tevalearningalliance.org

TorahTrek
Email: info@torahtrek.com
http://www.torahtrek.com

Resources for Judaism and the Environment Books

Abram, David. *The Spell of the Sensuous: Perception and Language in a More-Than-Human World.* New York: Pantheon, 1996.

Artson, Bradley Shavit. *It's a Mitzvah! Step-by-Step to Jewish Living.* New York: Rabbinical Assembly, 1995. See chapter 17 on Bal Tashkhit (prohibition of waste). "Leader's Guide" available.

Bach, David, with Hillary Rosner. *Go Green, Live Rich: 50 Simple Ways to Save the Earth.* NY: Broadway Books, 2008.

Benstein, Jeremy. *The Way Into Judaism and the Environment.* Woodstock, VT: Jewish Lights, 2006.

Bernstein, Ellen, ed. *Ecology and the Jewish Spirit: Where Nature and the Sacred Meet.* Woodstock, Vt.: Jewish Lights Publishing, 1998.

Bush, Lawrence and Jeffrey Dekro. *Jews, Money, and Social Responsibility: Developing a "Torah of Money" for Contemporary Life.* Philadelphia: The Shefa Fund, 1993. To order, call 1-800-92-SHEFA/215-247-9704.

Canfei Nesharim, *Compendium of Sources in Halacha and the Environment.* Jerusalem: Canfei Nesharim, 2005.

Carmell, Aryeh and Cyril Domb, eds. *Challenge: Torah Views on Science and Its Problems.* New York: Feldheim, 1976.

Conner, Nancy. *Living Green: The Missing Manual.* Sebastopol, CA: O'Reilly, 2009.

Conservative Judaism 44:1. Fall 1991. Entire issue dedicated to Jewish environmental thought and action.

Dorman, Josh. *The Lazy Environmentalist: Your Guide to Easy, Stylish Green Living*. NY: Stewart, Tabori & Chang. 2007.

E/The Environmental Magazine, Green Living. NY: Plume (Penguin), 2005.

Eisenberg, Evan. *The Ecology of Eden*. New York: Knopf, 1998.

Elon, Ari, Naomi Hyman and Arthur Waskow. *Trees, Earth and Torah: A Tu B'Shvat* Philadelphia: The Jewish Publication Society, 1999.

Foer, Jonathan Safran. *Eating Animals*. New York: Little Brown, 2009.

Goleman, Daniel. *Ecological Intelligence: The Hidden Impacts of What We Buy*. NY: Broadway Books. 2010

Gottlieb, Roger, ed. *This Sacred Earth: Religion, Nature, Environment*. New York: Routledge, 1996.

Heschel, A.J. *The Sabbath: Its Meaning for Modern Man*. New York: Farrar, Straus, Giroux, 1951.

Heschel, A.J. *God in Search of Man*. New York: Farrar, Straus, and Giroux, 1955.

Hill, Graham, and Meaghan O"Neill. *Ready, Set, Green: Eight Weeks to Modern Eco-Living*. NY: Villard, 2008.

Hill, Tessa. *The Everything Green Classroom Book*. Avon, MA: Adams Media, 2009.

Horn, Greg. *Living Green: A Practical Guide to Simple Sustainability*. Topanga, CA: Freedom Press. 2006.

Isaacs, Ronald H. *The Jewish Sourcebook on the Environment and Ecology*. Northvale, NJ: Jason Aronson, 1998.

Judaism and Ecology, a study guide produced by Hadassah and Shomrei Adamah. Send $12.00 to Dept. of Jewish Education, Hadassah, 50 W. 58th St., New York, NY 10019.

Kellert, Stephen, and Timothy Farnham, eds. *The Good in Nature and Humanity: Connecting Science, Religion, and Spirituality with the Natural World*. Washington, D.C.: Island Press, 2002.

Lieberman, Senator Joe. *The Gift of Rest: Rediscovering the Beauty of the Sabbath*. New York: Howard Books, 2011.

McDilda, Diane Gow. *365 Ways to Live Green*. Avon, MA: Adams Media, 2008.

McDilda, Diane Gow, *The Everything Green Living Book*, Adams Media, 2007

Oelschlaeger, Max. *Caring for Creation: An Ecumenical Approach to the Environmental Crisis*. New Haven: Yale University Press, 1994.

Rabinowitz, Louis I. *Torah and Flora*. Sanhedrin Press, 1979.

Riley, Trish. *The Complete Idiots' Guide to Green Living*. Indianapolis: Alpha Books, 2007.

Rockefeller, Steven C. and John N. Elder, eds. *Spirit and Nature: Why the Environment is a Religious Issue.* Boston: Beacon Press, 1992. See "Learning to Live with Less" by Ismar Schorsch.

Schwartz, Richard. *Judaism and Global Survival.* New York: Lantern Books, 2002.

Smith, Sharon J. *The Young Activist's Guide to Building a Green Movement and Changing the World.* Ten Speed Press. 2011.

Stein, David, ed. *A Garden of Choice Fruit: 200 Classic Jewish Quotes on Human Beings and the Environment.* Philadelphia: Shomrei Adamah, 1991. Available from Shomrei Adamah, 212.807.6376.

Tal, Alon. *Pollution in a Promised Land: An Environmental History of Israel.* Berkeley: University of California Press, 2002.

Tamari, Meir. *"With all Your Possessions": Jewish Ethics and Economic Life.* New York: Free Press, 1987.

Texts and Commentaries on Biological Diversity and Human Responsibility: A Study Guide (1996). Available from COEJL.

Tirosh-Samuelson, Hava, ed. *Judaism and Ecology: Created World and Revealed Word.* Cambridge: Harvard University Press, 2002. This book is part of a larger series called Religions of the World and Ecology, sponsored by the Center for the Study of World Religions at Harvard Divinity School; the series is edited by Mary Evelyn Tucker and John Grim.

To Till and To Tend: A Guide to Jewish Environmental Study and Action (1994, COEJL)

Waskow, Arthur. *Seasons of Our Joy: A Handbook of Jewish Festivals.* Boston: Beacon Press, 1982.

Waskow, Arthur. *Down to Earth Judaism: Food, Money, Sex, and the Rest of Life.* New Jersey: William Morrow, 1995.

Waskow, Arthur, ed. *Torah of the Earth: Exploring 4,000 Years of Ecology in Jewish Thought.* Woodstock, Vermont: Jewish Lights Publishing, 2000.

Wilzig, Tami Lehman. *Green Bible Stories for Children.* (Ages 8 to 11). Minneapolis: Kar-Ben Publishing, 2011.

Yaffe, Martin, ed. *Judaism and Environmental Ethics: A Reader.* Lanham, Md.: Lexington Books, 2001.

Yarrow, Joanna. *Eco Logical.* London: Duncan Baird Publishers, 2009.

Yarrow, Joanna. *1001 Ways to Save the Earth.* San Francisco: Chronicle Books, 2007.

Websites

Adamah Fellowship- http://www.isabellafreedman.orgladamah
Adam Teva ve'Din: The Israeli Union for Environmental Defense- http://www.iued.org.il
American Society for the Protection of Nature in Israel- http://www.aspni.org
Arava Institute for Environmental Studies- http://www.arava.org
Canfei Nesharim- http://www.canfeinesharim.org
The Coalition on the Environment and Jewish Life (COEJL)- http://www.coejl.org
EarthKosher- http://www.earthkosher.com
Eat Well – www.EatWell.com
US Environmental Protection Agency – www.epa.gov
The Eco-Kosher Network- http://www.ecojew.com/ecokashrut
Eden Village Camp - www.EdenVillageCamp.org
Elat Hayyim: The Jewish Retreat Center- http://jewishretreatcenter.org/overview/ecology_social_change.html
Friends of the Earth Middle East (FoEME)- http://www.foeme.org
Good Guide. Find healthy, green, ethical products according to scientific ratings. www.goodguide.com
The Green Guide Institute – www.thegreenguide.com
US Geological Survey (USGS) – www.usgs.gov
GreenHome.com – The Environmental Store - www.GreenHome.com
Green Zionist Alliance – www.greenzionism.org
Hazon- http://www.hazon.org
The Heschel Center for Environmental Learning and Leadership- http://www.heschelcenter.org/index_eng.html

Israeli Ministry of the Environment- http://www.sviva.gov.il (click on "English" in the upper left-hand corner)

Jewcology - a "one-stop shop" of Jewish-environmental resources shared by a wide range of Jewish environmental organizations – www.Jewcology.com

The Jewish Global Environmental Network (JGEN)- http://www.jgenisrael.org

Jewish National Fund- http://www.jnf.org

Jewish Vegetarians of North America- http://jewishveg.com

LetsGoGreen.biz - www.LetsGoGreen.biz

Life and Environment: The Israeli Union of Environmental NGO- http://www.sviva.net (click on "English" in the upper left-hand corner)

Local Harvest – www.localharvest.org

The Noah Project (London)- http://www.noahprojed.org.uk

Religious Action Center of Reform Judaism - http://rac.org/advocacy/issues/issueenv/

The Shalom Center- http://www.shalomctr.org

Society for the Protection of Nature in Israel - www.teva.org.il/english

Teva Learning Alliance – www.TevaLearningAlliance.org

Notes

Introduction

1. Berrata GP. World energy consumption and resources: An outlook for the rest of this century. International Journal of Environmental Technology and Management 2007; 7:99-112.
2. Organization for Economic Cooperation and Development/International Energy Agency. Key World Energy Statistics (OECD/IEA) 2008, Paris. 80 pp.
3. http://www.energybulletin.net/node/52312 ; http://www.fs.fed.us/pl/rpa/min89.htm
4. http://www.euwfd.com/html/source_of_pollution_-_overview.html
5. http://www.mfa.gov.il/MFA/Facts%20About%20Israel/Land/Israel-s%20Chronic%20Water%20Problem
6. De Villiers M. Water: The Fate of Our Most Precious Resource. Houghton Mifflin Company, New York, 2001, 352 pp.
7. op cit. OECD/IEA
8. United Nations Development Programme. World Energy Assessment Overview: 2004 Update. New York, 88 pp.
9. IPCC
10. An Assessment of the Intergovernmental Panel on Climate Change, approved in detail at IPCC Plenary XXVII (Valencia, Spain, 12-17 November 2007. These data are detailed in the Synthesis Report of over 2,500 scientists who gathered at the Intergovernmental Panel on Climate Change (IPCC) Plenary XXVII (Valencia, Spain, 12-17 November 2007)
11. For additional information refer to the US Climate Change Science Program Report Analyses of the Effects of Global Change on Human Health and Welfare and Human Systems (http://www.climatescience.gov/Library/sap/sap4-6/final-report/).

12. WGMS *Glacier Mass Balance Bulletin 2002-2003, No. 8.* World Glacier Monitoring Service, Zurich, 2005. http://www.wgms.ch/mbb.html
13. Cynthia Rosenzweig, Ana Iglesias, X.B. Yang, Paul R. Epstein, Eric Chivian. 2000. Climate Change and US Agriculture: The Impacts of Warming and Extreme Weather Events on Productivity, Plant Diseases, and Pests. Center for Health and the Global Environment, Harvard University. http://www.med.harvard.edu/chge
14. ibid.
15. IPCC
16. op cit IPCC
17. op cit IPCC
18. http://www.nationmaster.com/graph/ene_oil_con_percap-energy-oil-consumption-per-capita
19. 2005 Environmental Sustainability Index: Benchmarking National Environmental Stewardship, Yale Center for Environmental Law and Policy (Yale University) and Center for International Earth Science Information Network (Columbia University) In collaboration with: World Economic Forum (Geneva, Switzerland) and Joint Research Centre, European Commission (Ispra, Italy) http://www.yale.edu/esi/ESI2005_Main_Report.pdf
20. Primack, R. B. "Habitat destruction", Essentials of Conservation Biology. 4th Ed. Sinauer Associates, Sunderland, MA, 2006.pp. 177-188.
21. http://nationalzoo.si.edu/scbi/thinkgloballyactlocally/lossofhabitat/default.cfm
22. http://www.rawilsonfans.com/articles/CriticalPath.htm
23. An Assessment of the Intergovernmental Panel on Climate Change, approved in detail at IPCC Plenary XXVII (Valencia, Spain, 12-17 November 2007. These data are detailed in the Synthesis Report of over 2,500 scientists who gathered at the Intergovernmental Panel on Climate Change (IPCC) Plenary XXVII (Valencia, Spain, 12-17 November 2007)
24. For additional information refer to the US Climate Change Science Program Report Analyses of the Effects of Global Change on Human Health and Welfare and Human Systems (http://www.climatescience.gov/Library/sap/sap4-6/final-report/).
25. WGMS *Glacier Mass Balance Bulletin 2002-2003, No. 8.* World Glacier Monitoring Service, Zurich, 2005. http://www.wgms.ch/mbb.html
26. Cynthia Rosenzweig, Ana Iglesias, X.B. Yang, Paul R. Epstein, Eric Chivian. 2000. Climate Change and US Agriculture: The Impacts of Warming and Extreme Weather Events on Productivity, Plant Diseases, and Pests. Center for Health and the Global Environment, Harvard University. http://www.med.harvard.edu/chge
27. ibid.

28. W J Martens, L W Niessen, J Rotmans, T H Jetten, and A J McMichael. Potential impact of global climate change on malaria risk. Environ Health Perspect. 1995; 10:458-464.
29. Shriner DS, Street RB. North America. In: Watson RT, Zinyowera MC, Moss RH, editors. *The Regional Impacts of Climate Change. An Assessment of Vulnerability.* [A special report of the Intergovernmental Panel on Climate Change Working Group 2.] Cambridge (MA): Cambridge University Press; 1998. p. 253-330.
30. National Oceanic and Atmospheric Administration, National Climatic Data Center, http://www.ncdc.noaa.gov/oa/climate/globalwarming.html
31. D. King. Climate change: the science and the policy. Journal of Applied Ecology 2005; 42:779-783.
32. Board on Atmospheric Sciences and Climate, Committee on Global Change, National Research Council, *Chapter 3 Stratospheric Ozone Depletion: Global Processes,* Ozone Depletion, Greenhouse Gases, and Climate Change, National Academic Press, Washington, DC, 1989, pp. 10-18.
33. Hansen, J., et al. Dangerous human-made interference with climate: A GISS model study. *Atmos. Chem. Phys.*, 2007; 7:2287-2312.
34. IPCC op. cit.
35. ibid.
36. Ibid.
37. ibid.
38. 2004 UNESCO report, *Priorities for Research on the Ocean in a High-CO_2 World*
39. ibid.
40. F. Joos and R. Spahni. Rates of change in natural and anthropogenic radiative forcing over the past 22,000 years. Proc. Nat. Acad. Sci USA 2008; 105:1425-1430.
41. CLWeber and HS Matthews, Food-Miles and the Relative Climate Impacts of Food Choices in the United States, Environ. Sci. Technol., 42:3508-3513, 2008, http://pubs.acs.org/doi/full/10.1021/es702969f
42. Gamble, J.L., K.L. Ebi, F.G. Sussman, T.J. Wilbanks. 2008. Analyses of the Effects of Global Change on Human Health and Welfare and Human Systems. A Report by the U.S. Climate Change Science Program and the Subcommittee on Global Change Research. U.S. Environmental Protection Agency, Washington, DC, USA.
43. http://www.footprintnetwork.org/en/index.php/GFN/

Section A: The Basics

44. 43 global acres equals 0.07 square miles or 0.17 square kilometers.

[45] For example, in Wisconsin, the Department of Natural Resources along with the Milwaukee Public Museum and WE Energies (a public utility) have developed a program in which they have transplanted peregrine falcons into downtown Milwaukee. There is now an established colony of peregrines in the city with the added benefit that the local pigeon population is now under control. (http://images.library.wisc.edu/EcoNatRes/EFacs/PassPigeon/ppv49no04/reference/econatres.pp49n04.gsepton.pdf)

[46] See the text of this letter at http://earthrenewal.org/Open_letter_to_the_religious_.htm and further details at http://fore.research.yale.edu/publications/statements/joint_appeal.html

Section B: So Many Ways to Save Energy

[47] http://www.epa.gov/climatechange/basicinfo.html

[48] http://www.ucsusa.org/clean_energy/coalvswind/brief_coal.html
Where does this endnote belong??? US EPA Clean Energy. http://www.epa.gov/cleanenergy/energy-and-you/affect/air-emissions.html

[49] Union of Concerned Scientists. http://www.ucsusa.org/clean_energy/coalvswind/c02c.html

[50] US EPA eGRID. http://www.epa.gov/cleanenergy/energy-resources/egrid/index.html

[51] For a list of useful references see: http://en.wikipedia.org/wiki/Standby_power ; http://www.energystar.gov/index.cfm?fuseaction=find_a_product.showProductGroup&pgw_code=CO

[52] http://www.upenn.edu/computing/provider/docs/hardware/powerusage.html

[53] Bird L, Swezey B. US Department of Energy National Renewable Energy Laboratory, "Green Power Marketing in the United States: A Status Report (9th Ed.), NREL/TP-640-40904, November, 2006. http://www.osti.gov/bridge

[54] http://www.nature.com/nature/journal/v443/n7107/full/443019a.html

[55] According to documents posted at http://www.eia.doe.gov/iea/wecbtu.html, in 2006 the United States used of 40 Quadrillion Btu of Petroleum vs. World 172 Quadrillion Btus, and 23 Quadrillion Btu of Coal vs. World 123 Quadrillion Btu.

[56] See graphic representation, based on Department of Labor Consumer Expenditures, April 2009, at http://www.visualeconomics.com/wp-content/uploads/2009/07/wheredidthemoneygo.jpg.

[57] http://www.frost.com/prod/servlet/frost-home.pag

NOTES

58 http://www.heschel.org.il/eng/
59 Maimonides, Mishneh Torah, Mourning 14:24
60 Sefer Ha-Hinukh, Mitzvah #529

Section C: Bal Tashkhit: Reduce Waste

61 From "Bal Tashchit: Optimism in a Time of Teshuva" by Candace Nachman, on Canfei Nesharim'ss website at http://www.canfeinesharim.org/learning/torah.php?page=12439
62 http://www.coffeeresearch.org/market/usa.htm
63 http://www.edf.org/page.cfm?tagID=2155. For green dining best practices, see http://innovation.edf.org/page.cfm?tagID=30868.
64 http://news.nationalgeographic.com/news/2003/09/0902_030902_plasticbags.html
65 Woods Hole Oceanographic Institute. www.whoi.edu/science/B/people/kamaral/plasticsarticle.html
66 http://www.ctenvironment.org/PDFs/Decomposition%20rates%20chart.pdf
67 Will My Plastic Bag Still Be Here in 2507? How scientists figure out how long it takes your trash to decompose, June 27, 2007 at http://www.slate.com/id/2169287/
68 Horn G. Living Green: A Practical Guide to Simple Sustainability, Freedom Press, 2006, 171 pp.
69 http://www.epa.gov/epawaste/conserve/materials/ecycling/docs/fact7-08.pdf
70 http://www.epa.gov/epawaste/conserve/materials/ecycling/docs/fact7-08.pdf
71 US EPA. http://www.epa.gov/osw/conserve/materials/ecycling/manage.htm
72 http://ehp.niehs.nih.gov/members/2002/110-4/chart.jpg
73 http://ehp.niehs.nih.gov/members/2002/110-4/focus.html
74 To learn more, visit http://www.epa.gov/international/toxics/ewaste.html
75 http://www.usps.com/communications/newsroom/2008/pr08_028.htm
76 Beverage Marketing Corporation. U.S. and International Bottled Water Developments and Statistics for 2008 http://www.bottledwater.org/public/2008%20Market%20Report%20Findings%20reported%20in%20April%202009.pdf
77 See Message in a Bottle by Charles Fishman, at http://www.fastcompany.com/magazine/117/features-message-in-a-bottle.html
78 New York State Department of Environmental Conservation. http://www.dec.ny.gov/docs/materials_minerals_pdf/waterbottles.pdf
79 Public Health News Center, Johns Hopkins Bloomberg School of Public Health. 2008. http://www.jhsph.edu/publichealthnews/articles/2008/goldman_schwab_bpa.html
80 New York State Department of Environmental Conservation. http://www.dec.ny.gov/docs/materials_minerals_pdf/waterbottles.pdf
81 http://www.ewg.org/health/report/bottledwater-scorecard

82 Dasani: http://www.commondreams.org/headlines04/0304-04.htm; Aquafina: http://www.cnn.com/2007/HEALTH/07/27/pepsico.aquafina.reut/

83 These recommendations are from http://canfeinesharim.org/uploads/17390waterfacts.pdf written by Dr. Felicia Orah Rein for Canfei Nesharim.

Section D: Reduce, Reuse, Recycle

84 This report is available at http://www.news.cornell.edu/stories/March10/Gilovich-Happy.html.

85 http://www.wec.ufl.edu/extension/gc/harmony/documents/EH177.pdf

86 Maryland Department of the Environment. http://www.mde.state.md.us/programs/landprograms/recycling/education/process.asp

87 http://www.epa.gov/osw/nonhaz/municipal/pubs/msw2008rpt.pdf

88 http://www.epa.gov/osw/nonhaz/municipal/pubs/msw2008rpt.pdf

Section E: Buy Sustainably

89 http://www.cosmeticsdatabase.com/

90 This action was recommended in the dvar Torah, 'Where the Wood Meets the Water," by Rabbi Shmuel Simenowitz, available online at http://canfeinesharim.org/community/parshas.php?page=14317

91 According to Hallmark sales research, as cited at: http://corporate.hallmark.com/Product/Gift-Wrap-Overview

92 For a full listing of GDP by country, see http://data.worldbank.org/indicator/NY.GDP.MKTP.CD/countries/1W?cid=GPD_29&display=default

93 Unit For Sustainable Development and Environment. Organization of American States. http://www.un.org/esa/sustdev/natlinfo/indicators/idsd/themes/tourism.htm

94 http://www.ecotourism.org. See "Learning Center" for more on this topic.

95 http://www.mfa.gov.il/MFA/Facts+About+Israel/Looking+at+Israel/Looking+at+Israel-+Economy.htm

96 Union of Concerned Scientists. Getting There Greener The Guide to Your Lower-Carbon Vacation . 2008. http://www.ucsusa.org/assets/documents/clean_vehicles/greentravel_report.pdf

97 Cited at http://more.masortiworld.org/environment/space/home/Livinga_Responsible_Lifeand_KIs_FIVE_MITZVOT_FOR_SUSTAINABLE_LIVING.pdf

NOTES

Section F: Connect with Creation

98. "The Kabbalah of Clothes: Commentary on Tetzaveh," *Shraga's Weekly* by Rabbi Shraga Simmons http://www.aish.com/tp/b/sw/48950636.html
99. Quoted in Abraham Isaac Kook: The lights of penitence, the moral principles, lights of holiness, essays, letters and poems, published 1978.
100. "The Leaf," from the writings and talks of Rabbi Yosef Yitzchak Schneerson, online at http://www.chabad.org/liberary/article_cdo/aid/66990/jewish/The-Leaf.htm
101. Source unknown. As cited at http://more.masortiworld.org/environment/space/community/JEWISH_ENVIRONMENTAL_QUOTES.pdf, in a document created by the Teva Learning Alliance.
102. As quoted in *This Sacred Earth: Religion, Nature, Environment*, by Roger S. Gottlieb.
103. *The Brothers Karamazov*, p. 829.
104. *The Sense of Wonder*, 1956, republished January 1999.
105. "The Jewish Sabbath" in *Judaism Eternal* 30, as cited at http://canfeinesharim.org/community/parshas.php?page=14901

Section G: Greening Your Home

106. See my anthology, *A Shabbat Reader: Universe of Cosmic Joy*, URJ Press.
107. "Ecology in Jewish Law and Theology" in *Faith and Doubt: Studies in Traditional Jewish Thought*, Ktav Publishing House: Jersey City, New Jersey, 2006, p. 163-4
108. Ben Uziel 30, cited at http://canfeinesharim.org/community/parshas.php?page=14901
109. **Zota AR, Aschengrau A, Rudel RA, JG.** Self-reported chemicals exposure, beliefs about disease causation, and risk of breast cancer in the Cape Cod Breast Cancer and Environment Study: a case-control study. *Environmental Health* 2010, 9:40. doi:10.1186/1476-069X-9-40. http://www.ehjournal.net/content/9/1/40
110. Frumkin H. Safe and healthy school environments, Oxford University Press. UK.
111. See "Chemicals in Home a Big Smog Source", *LA Times*, March 9, 2003, http://articles.latimes.com/2003/mar/09/local/me-homesmog9
112. Klaassen CD. Casarett & Doull's Toxicology: The Basic Science of Poisons, 5th edition., McGraw Hill Co., Inc., 1996, 1111 pp.
113. http://www.nrdc.org/media/2007/070919.asp
114. http://www.epa.gov/WaterSense/water_efficiency/what_you_can_do.html

Section H: Sustainability and Jewish Eating

[115] Cunningham WP and Saigo BW. Environmental Science. WCB/McGraw-Hill, Boston. 1999. 650 pp.

[116] CDC (U.S. Centers for Disease Control and Prevention). 2009. Fourth National Report on Human Exposure to Environmental Chemicals. Department of Health and Human Services

[117] (http://www.foodnews.org/reduce.php)

[118] USDA. 2008. Pesticide Data Program: Annual Summary, Calendar Year 2008. United States Department of Agriculture. Agricultural Marketing Service. December 2009. http://www.ams.usda.gov/AMSv1.0/getfile?dDocName=STELPRDC5081750

[119] See http://blogs.consumerreports.org/baby/2008/06/organic-food.html, June 2, 2008.

[120] Learn about the USDA's National Organic Program (NOP) at http://www.ams.usda.gov/AMSv1.0/ams.fetchTemplateData.do?template=TemplateA&navID=NationalOrganicProgram&leftNav=NationalOrganicProgram&page=NOPUnderstandingOrganicLabeling&description=Understanding%20Organic%20Labeling&acct=nopgeninfo

[121] Learn more at http://www.hazon.org/go.php?q=/food/CSA/vision.html.

[122] The Talmud states that people were initially vegetarians: "Adam was not permitted meat for purposes of eating." (Sanhedrin 59b). Rashi (1040-1105), states the following about God's first dietary law: "God did not permit Adam and his wife to kill a creature and to eat its flesh. Only every green herb shall they all eat together." (on Genesis 1:29) Many other Torah commentators agree with this assessment, including Abraham Ibn Ezra (1092-1167), Maimonides (1135-1214), Nahmanides (1194-1270), and Rabbi Joseph Albo (died in 1444). Later scholars, such as Rabbi Samson Raphael Hirsch (1808-1888), Moses Cassuto (1883-1951), and Nehama Leibowitz (1905-1997), concur.

[123] Nahmanides, commentary on Genesis 1:29.

[124] By the time of Noah, humanity had degenerated greatly. "And God saw the earth, and behold it was corrupt; for all flesh had corrupted their way upon the earth" (Gen. 6:12). People had sunk so low that they would eat a limb torn from a living animal. As a concession to people's weakness, permission to eat meat was then given. Rav Kook wrote that the permission to eat meat was only a temporary concession, that a God who is merciful to His creatures would not institute an everlasting law permitting the killing of animals for food. See http://www.jewishvirtuallibrary.org/jsource/Judaism/ravkook_veg.html. As Rabbi Yitzhak Herzog, successor to Rav

Kook as Chief Rabbi of pre-state Israel (d. 1959) said, "Jews will move increasingly to vegetarianism out of their own deepening knowledge of what their tradition commands... A whole galaxy of central rabbinic and spiritual leaders...has been affirming vegetarianism as the ultimate meaning of Jewish moral teaching." (As cited at http://www.think-differently-about-sheep.com/Why_Animals_Matter_A%20 Religious_Philosophical_%20Perspective_%20Judaism_%20Quotations.htm)

125. FAOSTAT. (2001, 2002) FAO Statistical Databases (CD-ROM), Food and Agriculture Organization of the United Nations, Rome, Italy.

126. Schlosser E. Fast Food Nation: The Dark Side of the All-American Meal. Houghton Mifflin Co., New York, 2001. 356 pp.

127. Campbell TC and Campbell TM. The China Study: Startling Implications for Diet, Weight-Loss and Long-Term Health.BenBella Books, Inc., Dallas, TX. 2006. 422 pp.

128. For example: http://www.heart.org/HEARTORG/GettingHealthy/NutritionCenter/HealthyDietGoals/Fish-and-Omega-3-Fatty-Acids_UCM_303248_Article.jsp;

Koch RM, Jung HG, Crouse JD, Varel VH, Cundiff LV.Growth, digestive capability, carcass, and meat characteristics of Bison bison, Bos taurus, and Bos x Bison. J Anim Sci. 1995;73:1271-1281; Rule DC, Broughton KS, Shellito SM, Maiorano G.Comparison of muscle fatty acid profiles and cholesterol concentrations of bison, beef cattle, elk, and chicken. J Anim Sci. 2002;80:1202-1211;

Cynthia A Daley, Amber Abbott, Patrick S Doyle, Glenn A Nader and Stephanie Larson A review of fatty acid profiles and antioxidant content in grass-fed and grain-fed beef Nutrition Journal 2010, 9:10 doi:10.1186/1475-2891-9-10.

129. Zhao S, Zhang P, Melcer ME., et al. Estrogens in streams associated with a concentrated animal feeding operation in upstate New York, USA. Chemosphere 2010;79:420-425.

130. Heederik D, Sigsgaard T, Thorne PS, et al. Health effects of airborne exposures from concentrated animal feeding operations Environmental Health Perspectives 2007;115:298-302.

131. Bowman A, Mueller K, Smith M. Increased Animal Waste Production from Concentrated Animal Feeding Operations (CAFOs): Potential Implications for Public and Environmental Health. The Nebraska Center for Rural Health Research, Occasional Papers. Number 2, January, 2000. 17 pp.

132. See http://www.epa.gov/region7/water/cafo/index.htm

133. Understanding Concentrated Animal Feeding Operations and Their Impact on Communities, Centers for Disease Control, 2010, found at http://www.cdc.gov/nceh/ehs/Docs/Understanding_CAFOs_NALBOH.pdf

134. Saenz RA, Hethcote HW, Gray GC. Confined animal feeding operations as amplifiers of influenza. Vector Borne Zoonotic Disease 2006;6:338-346.
135. http://ehp03.niehs.nih.gov/article/fetchArticle.action?articleURI=info: doi/10.1289/ehp.117-a394
136. US Department of Agriculture, Agricultural Research Service, Manure and Byproduct Utilization: National Program Annual Report: FY 2001, www.nps.ars.usda.gov/programs/programs.htm?npnumber=206&docid=1076.
137. http://www.ncbi.nlm.nih.gov/pmc/articles/PMC1519243/pdf/envhper00360-0028-color.pdf
138. **Horrigan L, Lawrence RS, and Walker P.** How Sustainable Agriculture Can Address the Environmental and Human Health Harms of Industrial Agriculture. Environmental Health Perspectives Volume 110, Number 5, May 2002 110:445–456. http://ehpnet1.niehs.nih.gov/docs/2002/110p445-456horrigan/abstract.html
139. For a review of this topic and a list of references see: Putting Meat on the Table: Industrial Farm Animal Production in America. A Report of the Pew Commission on Industrial Farm Animal Production: A Project of The Pew Charitable Trusts and Johns Hopkins Bloomberg School of Public Health. 2008. 110 pp.
140. Gilchrist MJ, Greko C, Wallinga DB, Beran GW, Riley DG, and Thorne PS. The Potential Role of Concentrated Animal Feeding Operations in Infectious Disease Epidemics and Antibiotic Resistance. Environmental Health Perspectives 2007; 115: 313–316. Published online 2006 November 14. doi: 10.1289/ehp.8837.
141. Newsweek, March 28, 1994. See http://www.newsweek.com/1994/03/27/the-end-of-antibiotics.html, page 2 of 5.
142. See FDA draft guidance on antibiotic use on farm animals, in the Federal Register, June 29, 2010 http://edocket.access.gpo.gov/2010/pdf/2010-15289.pdf
143. National Corn Growers Association. http://www.ncga.com/livestock
144. http://www.smallplanet.org/books/frances-moore-lappe
145. *Diet for a Small Planet*, Frances Moore Lappe, p. 61. This book may be searched at http://www.amazon.com/Small-Planet-Frances-Moore-Lappe/dp/0345373669.
146. http://www.public-health.uiowa.edu/ehsrc/CAFOstudy/CAFO_6-2.pdf
147. Scientific Committee on Animal Health and Welfare, The Welfare of Cattle Kept for Beef Production, European Commission Health & Consumer Protection Directorate-General. SANCO.C.2./AH/R22/2000. 2001, 149 pp. http://ec.europa.eu/food/fs/sc/scah/out54_en.pdf
148. http://jcarrot.org/resources/kosher-sustainable-meat#resources
149. For more on this, see *The Omnivore's Dilemma* by Michael Pollan.
150. See a guide to the National Organic Program at http://helpguide.org/life/organic_foods_pesticides_gmo.htm

NOTES 233

151 For more on this, see *The Omnivore's Dilemma* by Michael Pollan.
152 Drawn from Canfei Nesharim's Tu b'Shevat speaker resources, direct citation at: http://canfeinesharim.org/uploads/11515speaking.pdf p. 6
153 http://www.grid.unep.ch/product/publication/download/ew_overfishing.en.pdf
154 "Impacts of Biodiversity Loss on Ocean Ecosystem Services", in *Science*, November 3, 2006, available with free registration at http://www.sciencemag.org/content/314/5800/787.full.
155 http://seagrant.gso.uri.edu/factsheets/Bycatch.html
156 http://dels-old.nas.edu/osb/pollution.pdf
157 http://www.environmentalhealthnews.org/ehs/news/ocean-mercury-increasing
158 See http://www.montereybayaquarium.org/cr/cr_seafoodwatch/sfw_health.aspx

Section I: Be a Green Jew

159 http://boulderjewishnews.org/2009/7-ways-to-green-a-barbat-mitzvah/
160 Among his other credentials, Rabbi Breitowitz is the former rabbi of the Woodside Synagogue Ahavas Torah in Silver Spring, MD.
161 "Reconnecting to Nature" by Rabbi Yitzchak Breitowitz, cited at http://canfeinesharim.org/community/shevat.php?page=11531
162 Full details about these facts are available at http://canfeinesharim.org/learning/environmental.php?page=21768
163 Full index of holiday resources from COEJL at http://www.coejl.org/~coejlor/celebrate/index.php
164 According to the view of Beit Hillel, see Mishnah Rosh Hashanah 1:1
165 US EPA http://www.epa.gov/heatisld/mitigation/trees.htm

Section J: Build a Green Jewish Community

166 Learn more at http://www.betterplace.com/
167 http://www.israel21c.org/briefs/israeli-campus-goes-green-cuts-costs
168 http://www.israelnationalnews.com/News/News.aspx/130565

DOV PERETZ ELKINS

Dov Peretz Elkins was born in Philadelphia. He is a graduate of Gratz College for Hebrew Teachers, received his BA in literature from Temple University, and his M.H.L. and rabbinic ordination from the Jewish Theological Seminary. He received his doctorate in counseling and humanistic education in 1976 at Colgate Rochester Divinity School. In 1989 he was given an honorary Doctor of Divinity degree for distinguished rabbinic service by his alma mater, the Jewish Theological Seminary.

After 2 years as military chaplain at Fort Gordon, Georgia, Rabbi Elkins served in several pulpits including Temple Beth El, Rochester, New York, one of America's largest and most prestigious congregations. From 1976 to 1985, he maintained a private practice in Pastoral Counseling, and was consultant to synagogues and many national Jewish organizations.

After serving as rabbi of Beth El Synagogue in Norfolk, VA, from 1987-1992 he was Senior Rabbi at the Park Synagogue in Cleveland, also one of the country's largest synagogues. In 1992 he became Senior Rabbi at The Jewish Center, Princeton, NJ. In Princeton he was President of the Clergy Association; created a computer learning center for the Religious School. He completed 13 years of service in Princeton at his retirement in July, 2005, and is now Rabbi Emeritus.

Dr. Elkins is a nationally known lecturer, educator, workshop leader, author, and book critic, he has written widely for the Jewish and general press, including such journals as *Reader's Digest, New Woman, The Christian Century, Judaism, Hadassah Magazine, Religious Education, Conservative Judaism,*

The Reconstructionist, and many others. Dr. Elkins has spoken on radio and television programs, and has been interviewed in cities all over the world for national and international media. He is a popular speaker on the Jewish circuit.

Rabbi Elkins is a recipient of the National Jewish Book Award, and is the author of over thirty-five books, including a two-volume collection of inspirational readings on the High Holidays, *Rosh Hashanah Readings: Inspiration, Information and Contemplation* and *Yom Kippur Readings* (Jewish Lights Publishing).

His *Chicken Soup For The Jewish Soul*, co-edited with Jack Canfield and Mark Victor Hansen (Health Communications, September, 2001) was on the NY Times best-seller list. Among Rabbi Elkins' other recent books is *The Bible's Top Fifty Ideas: The Essential Concepts Everyone Should Know*, and *The Wisdom of Judaism: An Introduction to the Values of the Talmud*, Jewish Lights Publishing. His most recent books are *Jewish Stories from Heaven and Earth: Inspiring Tales to Nourish the Heart and Soul* (Jewish Lights) and *Tales of the Righteous* (Compiled by Simcha Raz, translated by Dov Peretz Elkins, Gefen Publishing).

Dov Peretz Elkins is on the Board of Regents of Gratz College, Philadelphia, where he was instrumental in developing an international program of on-line courses for teens.

Rabbi Elkins and his wife, Maxine, reside in Princeton, NJ. Their six children and nine grandchildren live in Los Angeles, Israel, Cleveland, and New York.

His web sites are www.JewishGrowth.org, and www.Eco-Judaism.org. His email is RabbiElkins@gmail.com.

CPSIA information can be obtained at www.ICGtesting.com
Printed in the USA
BVOW040853191212

308678BV00002B/436/P